FIRST MAN

THE ANNOTATED SCREENPLAY

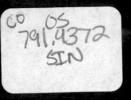

FIRST MAN
THE ANNOTATED SCREENPLAY
ISBN: 9781785659997

Published by
Titan Books
A division of Titan Publishing Group Ltd
144 Southwark Street
London
SE1 0UP

www.titanbooks.com

First edition: October 2018

10 9 8 7 6 5 4 3 2 1

Did you enjoy this book? We love to hear from our readers.
Please e-mail us at: readerfeedback@titanemail.com or write to
Reader Feedback at the above address.

To receive advance information, news, competitions, and exclusive offers online, please sign
up for the Titan newsletter on our website: www.titanbooks.com

A CIP catalogue record for this title is available from the British Library.

Printed and bound in Canada.

Publisher's note: The screenplay included in this book is the correct version as of our press date.

Afterword by

RICK ARMSTRONG AND MARK ARMSTRONG

TITAN BOOKS

FOREWORD

DAMIEN CHAZELLE

To be completely honest, I didn't know much about Neil Armstrong or the Moon mission before starting to work on *First Man*. But I was intrigued. And once I started digging in, I found myself fascinated by the psychology of the real men behind the mythology. I found myself asking, what does it take to strap yourself into one of those rickety tin can capsules and get hurtled into the unknown? I have a hard enough time surviving a five-hour transatlantic flight. Once I got into the details of Gemini and Apollo, I couldn't stop myself from marveling at the audacity, the unlikeliness, and the near-madness of it all.

Most of the cinematic representations of the Space Race I saw growing up were exciting and celebratory. But one thing I wanted to capture even more directly was a visceral sense of what it would actually feel like to be in one of those tiny command modules—an ordinary, fragile human being with only the thinnest layer separating oneself from the awesome terror and grandeur of space. And this begged another question... How do you match the cosmic to the individual and intimate?

I wanted to capture these conflicting ideas by shooting the film like a documentary, which meant that it had to be written like one. It turns out this is much harder than it looks. There had to be a total naturalism to the dialogue, while also being technically accurate. The scenes had to unfold as if they hadn't been planned—this while maintaining a clear structure.

And we'd have to pack an enormity of famous incidents into a short amount of time without feeling rushed. It became clear quite early on that the demands on the script level would be enormous.

I needed a writer who could match the journalistic with the artistic. *Spotlight* hadn't come out yet, but from my first meeting with Josh I could sense that this was his passion. Sure enough, Josh became my go-to, my guide, my collaborator, and my partner. I had never made a film I hadn't written before, so all of this was new to me, and it was exhilarating. We bounced ideas off of each other, saw what worked and what didn't, and Josh kept the project alive and churning during the time I was shooting *La La Land*. Josh is peerless in the way he is able to match history with the personal, and together we became obsessed by this story, by the cost of taking on such an extraordinary mission—not just the financial cost, but also the physical, emotional, and psychological costs on the individuals in the space program.

Working with Josh has been one of the greatest experiences of my professional life; our trip to the Moon and back was thrilling. What I love about these pages is that they offer people an opportunity to see what that expedition entailed and the hard choices we made in trying to bring Neil's experience to a new generation. I hope you enjoy the journey as much as I have.

ONE GIANT LEAP:
CREATING A HISTORICAL MOTION PICTURE

JOSH SINGER

How do you write a movie about Neil Armstrong? The man was a hero and an icon. Revered in the space community. Beloved in the greater world. He's one of the more famous figures in American history. Maybe in all of history. How do you write a movie about that guy? The short answer is, *you don't*. You don't write a movie about an icon. You try to write a movie about a man, the real human being who lived that life, who achieved those things. You try to put yourself in *his* shoes, experience what *he* experienced, feel what *he* felt.

This was our approach. I say "our" because, from the very beginning, this was a collaboration. Sure, I put pen to paper. But, from the outset, I was working incredibly closely with two men who I consider myself fortunate to call friends.

Damien Chazelle had a clear-eyed vision for this film from the start. We broke the pitch and started the research together. Then he went off to shoot a little movie musical and I spent a few months putting together a little outline. Okay, it was a seventy-five-page outline. It's kind of amazing anyone managed to get through it, let alone a director knee deep in prep. But Damien did it. He gave me a round of notes and I went back to work on the first of what would be hundreds of drafts. Along the way, we got input from all sides. Our producers Isaac Klausner, Wyck Godfrey, and Marty Bowen all had wonderful thoughts, as did the actors—particularly Ryan who, not surprisingly, has an incredible instinct for character. But it was Damien's laser sharp story focus that was my true north throughout the writing process.

If Damien provided the compass, James R. Hansen provided the map. As a professor of History, Jim has devoted much of his life to studying and writing about the history of NASA. His book on Neil Armstrong is both definitive and encyclopedic and, for a professional dramatist, a goldmine of in-depth research and surprising insights. Jim's help and

support throughout the creative process was invaluable. He read every draft and gave me hundreds of invaluable notes along the way. He also introduced us to all manner of NASA folks who helped make the script (and then the movie) as real as possible. But as large as that contribution was, it's tiny in comparison to the foundation Jim gave us. Surely, this movie would not and could not exist without Jim's incredible research.

- Neil lost his daughter mere months before he applied to be an astronaut.
- Neil's boss at NASA did not support his astronaut application.
- Neil's close friend Elliot See died in a plane crash two weeks before Neil's first space mission.

For us, these facts, which seem to have gone largely unnoticed by the wider public, helped shed light on the man behind the myth. As did conversations with Neil's family, friends, and colleagues. All of which only gave us an even greater appreciation for what Neil accomplished.

And this was our goal. To get at who Neil was and what it was like to do what he did, to share this history with the world through the canvas of a motion picture—this was our primary goal. But it wasn't our only goal. As dramatists, we had another mandate: Tell a good story. Tell a good story in two hours.

Movies are entertainment. If they aren't entertaining, nobody shows up. And if nobody shows up, it doesn't matter how interesting or thought provoking your historical thesis is. You can't get people to engage if you don't get them into the theater.

These two goals, good story and good history, aren't mutually exclusive. Neil's first space mission, Gemini VIII, is a great story. As is the first lunar landing and Neil's LLTV ejection and Neil's X-15 flight at the start of our film. But to put them in context of a greater story, a story about loss and sacrifice and, yes, incredible heroism, one invariably has to make compromises.

Temporal fictions are, to me, the least egregious. The X-15 flight we open the film with happened *after* Neil's daughter Karen died, not before. But the point is the same—a number of people, including Neil, thought there was a connection between her death and his performance. So, a temporal shift that lets us meet Karen while she's alive seems reasonable.

Fictions of place also seem to me to be mostly fair game. No, Deke probably didn't assign Neil to Gemini VIII right on the Gemini V launchpad. But, this helps move us through our story, provides a great setting and allows us to have Elliot See in the background when Neil learns he's going to be separated from his Gemini V (backup) pilot.

For my money, fictions of manner are by far the most delicate. Did the Gemini astronauts train on the Multi-Axis Trainer? Did Neil break a glass when Ed White died? Did Janet really yell at Neil when she pushed him to talk to Rick and Mark before he left for Apollo 11? No, probably not. But it was only after a lot of thought and debate with Jim and any number of other experts (including Frank Hughes, Al Worden, Gerry Griffin, Charlie Duke, Dave Scott, Mike Collins, Buzz Aldrin, and the Armstrong family) that we decided to include these fictions in our movie.

Which brings me to the point of this book. I think it's fair to say that all of us involved in the production of *First Man* felt a tremendous responsibility not just to tell a good story but also to get the history right. Or at least as right as possible within the framework of a modern motion picture. We were very careful, cautious even, when we chose to bend that history to fit the frame of our story. Which is why, to avoid any confusion, we want to be clear about where we did that, and why.

In the pages that follow, you'll find the *First Man* script with an accompanying conversation between me and Jim Hansen. We try to give some greater historical context for various key moments in the script. We also try to separate those facts from the dramatic fictions.

It is our hope that by being transparent about the choices we've made and why we made them, we'll both avoid historical confusion and give some insight into our thinking.

Of course, we also hope that this dialogue will provide an interesting means for historians, cinephiles, and general moviegoers to more deeply engage with Neil Armstrong. And Ed White, Elliot See, Bob Gilruth, Chris Kraft, Deke Slayton, Dave Scott, Buzz Aldrin, Mike Collins, and the entire NASA team. Not to mention, Janet and Rick and Mark Armstrong, Pat White, and Marilyn See. All these folks are heroes. And, as I said above, I think we appreciate their heroism on a deeper level now that we've gotten to know all of them a little bit better as people. We hope you feel that way too.

"WE ALL LIKE TO BE RECOGNIZED NOT FOR ONE PIECE OF FIREWORKS BUT FOR THE LEDGER OF OUR DAILY WORK."

NEIL ARMSTRONG

FOR THE ARMSTRONGS.
AND ALL THE FAMILIES THAT
SACRIFICED TO GET US THERE.

First Man

Based upon the book
First Man by James R. Hansen

*[**NOTE:** This script reflects where the edit is three weeks prior to picture lock. We are still refining the cut, adjusting ADR, etc., so there may be some discrepancies with the finished film.*

To further demonstrate how things change in the editing room, the script includes several scenes/lines that won't appear in the final film. For the most part, this is noted accordingly.]

WRITER'S OUTLINE - 07/22/15
WRITER'S DRAFT - 09/24/15
WRITER'S REVISED - 10/22/15
WRITER'S 2ND REVISED - 12/15/15
WRITER'S 3RD REVISED - 01/26/16
FIRST STUDIO DRAFT - 02/26/15
WRITER'S 5TH REVISED - 03/19/16
WRITER'S 6TH REVISED - 04/08/16
2ND STUDIO DRAFT - 04/23/16
WRITER'S 8TH REVISED - 05/13/16
3RD STUDIO DRAFT - 07/08/16
WRITER'S 10TH REVISED - 08/06/16
4TH STUDIO DRAFT - 08/25/16
WRITER'S 12TH REVISED - 09/11/16
WRITER'S 13TH REVISED - 12/05/16
WRITER'S 14TH REVISED - 01/14/17
WRITER'S 15TH REVISED - 03/22/17
WRITER'S 16TH REVISED - 06/24/17
PREP WHITE - 07/21/17
PREP BLUE - 09/17/17
WHITE SHOOTING DRAFT - 09/18/17
BLUE REVISIONS - 10/05/17
PINK REVISIONS - 10/27/17
YELLOW REVISIONS - 11/06/17
GREEN REVISIONS - 11/17/17
GOLDENROD REVISIONS - 12/14/17
BUFF REVISIONS - 12/17/17
SALMON REVISIONS - 01/11/18
CHERRY REVISIONS - 01/14/18
TAN REVISIONS - 01/20/18
2ND BLUE REVISIONS - 01/27/18
2ND PINK REVISIONS - 02/02/18
POST CONFORMED WHITE - 5/22/18
POST CONFORMED BLUE (IN PROCESS) - 6/29/18

OVER BLACK:

We hear a **LOW RUMBLE.**

It gets louder as we hear... a SCREAMING **ENGINE**... HOWLING **WIND**... BURSTS of **STATIC**... and FAINT **COMMS.**

It SURROUNDS us, filling us with dread, **POUNDING US INTO --**

1   <u>**INT. X-15 COCKPIT, HIGH RANGE, ABOVE EDWARDS AFB - DAY**</u>   1

<u>A pair of **BLUE EYES.**</u> **TICKING** back and forth. Rapidly. Ignoring the FRIGHTENING WALL OF SOUND all around us.

> JOE (COMMS)
> *Data check?*

> NEIL (O.C.)
> 2 APU on. Cabin pressure is good, 3500 on #1, 3355 on #2. Platform internal power.

PULL BACK TO **NEIL ARMSTRONG**, 31, in a silver pressure suit. Neil is INTENSELY FOCUSED, impressive in the SEVERE TURBULENCE.

> JOE (COMMS)
> *What's your mixing chambers?*

> NEIL (INTO COMMS)
> -44 and -45.

> BUTCH (COMMS)
> *Two minute point.*

> NEIL (INTO COMMS)
> MH circuit breakers on, opening nitrogren valve.

Neil opens the nitrogen valve on the low tech console. As the nitrogen creates a THIN WHITE FOG in the cabin, Neil looks out the window. *The plane looks like a **ROCKET**... <u>because it is</u>.*

<u>**This is the X-15**...</u> the <u>FASTEST FUCKING AIRCRAFT EVER MADE</u>. Hence the nitrogen. Neil shivers a bit.

> NEIL
> Chilly.

> JOE WALKER (COMMS)
> Won't be for long.

We note the X-15 isn't flying exactly, it's under the wing of a B-52, an **EIGHT ENGINE BEHEMOTH** shaking more than the X-15. TERRIFYING, but Neil's calm as he's **KNOCKED** about the cockpit. Neil closes the nitrogen valve.

> NEIL (INTO COMMS)
> Precool is off. Little bumpy.

<u>Classic Armstrong understatement</u>; underscored over comms by the pilot of the B-52 (BUTCH).

JOSH: So if this is a Moon mission movie why start with the X-15? The short answer is that we fell in love with the aircraft. The fastest and highest flying aircraft ever built, the X-15 flew well over Mach 6 (4,520 miles per hour) and more than 50 miles high, well outside the sensible atmosphere. Plus airplanes were Neil's first love.

JIM: That's true. He really didn't enjoy talking about the Moon landing, probably because that was all anyone ever asked him about. But ask him about the North American X-15 and he'd talk a blue streak.

JOSH: Not to mention that Damien's approach to filming the flight from Neil's point of view gives you a sense of how technically skilled all the X-15 pilots had to be. There weren't many of them, Neil was flying in some pretty rarefied air.

ABOVE: Ryan Gosling stands in front of the X-15 mock-up, evoking the historic photo of Neil Armstrong.
RIGHT: (Top) Neil Armstrong photographed in 1960 following a mission in the first X-15 rocket plane. (Bottom) Major-General Joe Engle was a technical consultant on the film. He is a former astronaut and the last living X-15 pilot.

JOSH: I traveled up to the Armstrong Flight Research Center at Edwards Air Force Base where Neil had flown his X-15 missions. Gene Matranga, one of the research engineers who worked with Neil, walked me through the entire flight. And the NASA rep Cam Martin gave me copies of the flight transcript from this flight, as well as Neil's pilot comments and pilot notes.

JIM: What is great about the script here (and on all of Neil's flights in the movie) is that the dialogue uses the actual words and phrasing that Neil and his fellow pilots and ground crews actually said.

JOSH: "Little bumpy" is one of my favorites. Very characteristic, the way he plays down a stressful situation. And it gives you an idea of how much trouble he was in when he admits later that he's "in pretty bad shape for the south lake bed."

JIM: You don't use the entire transcript and it isn't everything that they said, but it's the essential communication we need to hear to understand what's going on.

JOSH: In postproduction, we decided we wanted the dialogue to be even more spare so you could experience the ride. We wound up losing everything that was unnecessary, including most of Neil's comms.

JOSH: I've left the dialogue here (with cross outs to indicate what's been cut from the movie) so readers could get a more granular sense of what happened on this flight, and what exactly distracted Neil. What doesn't come from the transcript comes directly from Neil's post-flight notes and pilot comments. Neil's reports were incredibly helpful.

> BUTCH (COMMS)
> *Worst it's ever been, real rough up*
> *here, fluctuating a half degree each*
> *side.*

But this just drives Neil into deeper focus. His eyes **TICK
METHODICALLY** from gauge to gauge.

> NEIL (INTO COMMS)
> Velocity 900 fps, altitude 44,500,
> igniter ready to light.

BUTCH (COMMS) *Twenty seconds to drop --*	NEIL (COMMS) Rog, precool on, lox pump bearing plus eight.

In the bumpy cockpit, Neil grabs the stick.

JOE (COMMS) *Arm switch lite checks.*	NEIL (INTO COMMS) Going to prime. Ammonia up.

BUTCH (COMMS) (CONT'D) *Ten seconds --*	NEIL (INTO COMMS) Igniter idle, ready to launch on 3, 2, 1, release --

The X-15 is **RELEASED** from the wing of the B-52. Through the
window, we see the B-52 **RISE AWAY** as the X-15 **DROPS IN FREE
FALL**... *A fall we feel in our gut because...*

WE'RE NOT CUTTING AWAY. *We're gonna be IN THE COCKPIT with
Neil for this ENTIRE HEART-POUNDING RIDE.*

So, we're **DROPPING**. **_FAST_**. Neil pulls the throttle inboard.
The rocket LIGHTS, WE HEAR THE ROAR OF 57,000 LBS OF THRUST.

The rocket **TAKES OFF**. Neil's **PRESSED** into his seat.

> JOE (FAINT, COMMS)
> *Good on track, 15 seconds.*

Neil pulls on the stick. The plane **RISES SHARPLY**... it's
black nickel nose glowing **CHERRY RED** from the heat. The **GLOW**
lights up the cockpit, extending around Neil's legs and lap.

Neil starts to sweat in the heat. Heavily. Sweet Jesus.

Neil's eyes **TICK** to the analogue altitude gauge, SPINNING UP
past 150,000 feet. BLUE SKY NOW TURNING TO BLACK.

> NEIL (INTO COMMS)
> I'm indicating 5,800, pushing over.

We're over **MACH 5** and we feel it... until Neil CUTS the engine, **JOLTING** him against his harness.

All is STILL. Quiet. The radio BUZZES, but it's far away.

> NEIL (INTO COMMS)
> Top view, looking out, can see an
> awful long ways.

A **PENCIL FLOATS PAST NEIL** and we realize... we're suborbital. So yeah, things float. Neil's eyes move from the book to the earth curving away below the **BLACK SKY**. The STARS. The Moon.

He pauses, breathes it in with BOYISH AWE... and a **REVERENCE**. The world below seems distant and serene.

We hear a GARBLED **BUZZ** over comms.

> JOE (COMMS)
> Twenty degree angle of attack, check
> your Gs.

Neil notes the G-force gauge (**G-GAGE**) climbing up to **3.5 G**... He **FOCUSES** on the G-gage. A beat, then...

> JOE (COMMS)
> Okay, 140,000 feet, on your way
> down.

The plane descends, starts to **SHAKES** as the altitude gauge spins down through 135,000, **BLACK SKY FADING AGAIN TO BLUE**...

> JOE (FAINT, COMMS)
> ...approaching 115,000 feet, should
> be regaining aerodynamic controls...

Little more shake. Neil reaches for the stick and focusing again on the G-gage, now at **4 Gs**.

> JOE (FAINT, COMMS)
> Okay..* right turn.

WIND **WHIPS** over the plane. Neil pulls the stick right, <u>but</u>
<u>his eyes remain focused on the G-gage</u>...

> NEIL (INTO COMMS)
> Sixty degree bank. Climbing past 4
> Gs, the G-limiter should kick in...

...until he notices the altitude gauge <u>SPINNING UP</u>. *115, 116,*
117... the sky turns from blue back to black.

The shaking stops and the wind fades to eerie silence...

> JOE (COMMS)
> *...we show you ballooning, not*
> *turning. Hard right turn.*

Neil's eyes **TICK** to the windows, he **PULLS** on the stick...

> JOE (GARBLED COMMS)
> *...altitude rising... lot more*
> *right.*

Neil hits a few buttons on the console...

> NEIL (INTO COMMS)
> Didn't appreciate the altitude I was
> at, increasing deflection on the
> stabilizers. Should be enough air
> to bite into...

Neil pulls at the stick again, his eyes **TICKING** from his
gauges to the vista racing by despite his maneuver...

> JOE (COMMS)
> *Neil, you're bouncing off the*
> *atmosphere...*

The altitude gauge spins up through 120,000 feet.

Neil struggles with the stick. He remains calm, but through
the windows we still see BLACK SKY, the world flying by; *it's*
like he's drifting off the face of the planet.

> JOE (COMMS)
> *Altitude still rising, hard right!*

The speed at which the world passes, the lack of control, it's
terrifying... but Neil just works the problem. He eyes the
altitude, climbing past 140,000 feet... <u>realizing something</u>.

> NEIL
> Air's too thin, aerodynamic controls
> not responding, switching to
> reaction controls.

JIM: Neil elected to leave his angle of attack at about 15 or 16° because he was hoping to see the new g limiter in action. He had seen g limiting on the X-15 simulator at 4 to 4.5 g's and he was hoping to trigger it here. That's where he got into the ballooning situation.

JOSH: This happened somewhere between 85–120,000 feet. The flight path map indicates the height of the bounce was between 20–30,000 feet. To emphasize how dangerous this was we've used the high end of that range. And in the script we used Neil's pilot comments to help explain what happened. Things like "maintaining angle of attack to engage the g limiter" and "I didn't properly appreciate the altitude I was at," are pretty much verbatim from Neil's pilot comments. I ran lines like these by Joe Engle, the last surviving X-15 pilot. Joe checked to make sure everything sounded authentic and was true to what had happened on the flight. Of course, here the film is closer to reality as we've lost most of this dialogue.

JIM: Joe also reviewed the original comms and caught a couple of mistakes they must have made in the transcription.

JOSH: Joe was tremendously helpful with all aspects of this sequence. He not only walked me through the flight several times, but he worked with Nathan Crowley, our production designer, and trained Ryan and our insert unit on how to fly the plane. We had him on set for almost the entire X-15 shoot. And he reviewed the movie in post.

JIM: In the movie the X-15 passes through some pretty heavy cloud cover over Edwards.

JOSH: Yeah, we needed the clouds to show just how fast the X-15 was flying. But they probably wouldn't have flown on such a cloudy day. Joe was pretty funny when he saw them.

JIM: Yes, I recall him ribbing you there. But Damien did need a visual reference for how fast the X-15 was flying. And in your defense, Stan Butchart did say it was the worst turbulence he'd ever seen.

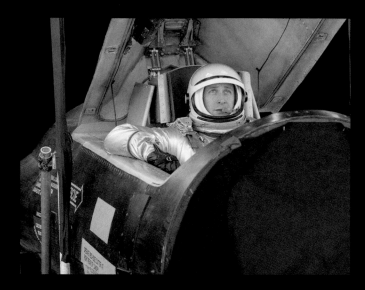

TOP: Ryan Gosling in the X-15 mock-up getting ready to shoot.
MIDDLE: The X-15 mock-up was mounted on a gimbal rig. The 35 by 60 foot 180° surround LED screen projected the flight as the gimbal rig shook the plane—this tends to get a more tactile image than standard green screen.
ABOVE: Ryan Gosling training on the X-15 console, under the watchful eye of Joe Engle and second assistant director Spencer Taylor.

He drops the stick, squeezes the ballistic controls. A burst
of gas shoots down from the left wing...

TOSSING the left of the plane **UP TOWARDS THE SKY**...

Black sky to Neil's left, the earth to his right; *Neil's
banked into a 90 degree turn...*

...but out the window, the world is still slipping by.

Neil HITS the ballistic controls again... a BURST of gas, this
time from the plane's nose, pushing the plane's nose DOWN.

> NEIL (INTO COMMS)
> Decreasing angle of attack to
> increase airspeed...

Neil eyes the **ALTITUDE GAUGE**. *145, 146...* **HOLDING** at *147,000
FEET...* and then starting to fall. *146, 144, 141....*

> NEIL (INTO COMMS)
> Coming down now...

...the plane starts falling **FASTER**. *100, 95, 90...*

G-forces **PRESS** Neil into his seat as the sky **FADES TO BLUE**.
The vista stops sailing by, the plane finally starts to turn.

The plane **GROANS** under the strain of the steep bank and Neil
eyes the gauges, reaching for the stick...

> NEIL (INTO COMMS)
> 350 knots, switching back to
> aerodynamic --

Neil's **SMASHED into his seat;** we feel every bit of the steep
bank as aerodynamics take hold and we start turning right, the
vista moving back the other way. We're also FALLING. **FAST.**

> NEIL (INT COMMS)
> ...surfaces bottoming out...

75, 70, 75... Neil wrestles with the stick as we bank and **FALL**
at **TERRIFYING SPEED**...

At last, he **PULLS** the lifeless plane **LEVEL**. *Jesus.*

> JOE (COMMS)
> *You seem to be in position.*

> NEIL (INTO COMMS)
> (scans his gauges)
> I agree, 300 knots, coming down now.

                    JOE (COMMS)
          Can you give us an estimate of your
          location?

Neil eyes tick from his gauges to the window, taking stock...

                    NEIL (INTO COMMS)
          Looks like I'm pretty, in pretty bad
          shape for the south lake bed.

                    JOE (COMMS)
          Okay, working the contingencies for
          a landing from the south.

Neil scans the landscape for the desert runway as the plane
DIVES LIKE A BRICK towards the mountains. But he knows...

                    JOE (COMMS)
          Neil, there is no contingency. You
          need to get back to Rogers.

40, 35, 30... Neil **BEARS DOWN**, WILLING the engine-less plane
in. The plane **SHIMMIES**, wind **BUFFETING HIM** at 250 knots.

          JOE (COMMS)   NEIL (INTO COMMS)
8 degree angle.   Affirmative. I'm going to
                                   jettison now.

He flicks a switch; a hiss as oxygen and ammonia jettison out
the rear the rear of the plane.

                    NEIL (INTO COMMS)
          I can see the base, the landing will
          be on south lake.

His eyes **DART** from the distant landing strip to the altitude
gauge, dropping 20,000 feet per minute. He closes the speed
brake handle... **WHITE GAS POURS** from the instrument panel.

                    JOE (COMMS)
          You're gonna have to make in a
          straight in approach. You'll have
          to stretch out your glide.

The cockpit is **FILLED** with the WHITE NITROGEN GAS FOG... The
gas slowly dissipates, but the view is HARDLY A RELIEF.

          JOE (COMMS)
You seem to be a tad short.   NEIL (INTO COMMS)
                                   I'm a little shorter than I
                                   thought.

JOSH: We do take license with a few other things. Neil banks right instead of left—this was a visual decision related to the flight path and the trajectory of the Sun. And Neil barely clears the mountains when in fact he was well above them. In reality it was a field of Joshua trees he struggled to clear, which was no less dangerous, but we felt the mountains better conveyed the potential jeopardy of Neil's straight in approach.

JIM: We should talk about the choice to open a movie about Neil Armstrong with a sequence in which he seemingly makes a mistake in his piloting. Why make that choice?

JOSH: Well, it does make for a great opening set piece. But there was a deeper logic here. One of the goals of this film is to humanize Neil. And this was one of three such flying incidents that occurred in the wake of the death of Neil's daughter, Karen.

JIM: Most people don't know that Neil lost his two-year-old daughter Karen to a brain tumor in January 1962, nine months before he became an astronaut and joined Project Gemini in Houston.

ABOVE: Brian d'Arcy James and other crew members prep to shoot at Edwards Air Force Base.
BELOW: The tail of the X-15 mock-up was dragged along the desert to achieve this shot.

JOSH: In your book you look very carefully into the issue of whether Neil's flying issues in that time period were related to Karen's death.

JIM: Even Neil later said, "I'd have to think that [my] performance was somewhat affected by the situation."

JOSH: The pain of that loss—a loss that was deep enough to affect his performance—we felt that was one key to understanding the man behind the myth.

JIM: That leads to another question. You play with time here. Neil did have an X-15 flight in December 1961, before Karen died. But the flight you depict (and the other two mishaps he had) occurred in April/May 1962—two months after Karen's death. Why make that change?

JOSH: We wanted to meet Karen in person. To show him with her so an audience could feel how devastating her death was for Neil. So, yes, we take license and suggest this flight happened just before her death, rather than just after. But we felt the proximity of these flying issues to her death was more important than the exact timeline.

JIM: Given how emotionally difficult life became for Neil and Janet as soon as Karen's health situation surfaced, it certainly seems appropriate to me as his biographer to portray Neil's troubled performance in the air as coming during her illness rather than after her death.

First Man   POST CONFORMED BLUE   7.

Neil pulls the stick to hold altitude, but he keeps **DROPPING**.
9,000... 8,500... 8,000 feet... <u>Fuck</u>. It's gonna be close.

We see a CHASE PLANE **SWING UP** on Neil's right.

              CHASE PILOT (COMMS)
Neil, you can punch your   NEIL
ventral...   Okay, shoulda done that
                              sooner.

Neil, <u>ANNOYED</u> at himself, **HITS** the ventral jettison button.
The ventral fin **BLOWS OFF** and the plane **LURCHES** forward...
Neil **STRUGGLES** with the stick... *4,000, 3,500, 3,000 feet...*

              CHASE PILOT (COMMS)
Start your flaps down now.   NEIL
                              Thank you.

Neil **LOWERS** the flaps. The plane slows a bit more...

PUSH IN on Neil, INTENSITY **BURNING**. *2,000, 1,000, 500 feet...*
Neil **HURTLES** past the brush, *JUST CLEARING THE JOSHUA TREES!*

              CHASE PILOT (COMMS)
       You're in! Go 'head, put her down!

Neil **PULLS UP,** flaring the plane... It SLAMS DOWN with a **BANG,**
SKIDS ROUGHLY across the lakebed... **SHAKING VIOLENTLY...**

Neil **OPENS** the back fin brakes and the plane SWERVES, **TOSSING**
Neil and **KICKING UP** a HUGE CLOUD of dust... until at last, it
eases to a **HALT**.

For a moment, all is STILL. Silent. Then Neil STIRS...

              NEIL (INTO COMMS)
       I'm down.

       CHASE PILOT (COMMS)   JOE (COMMS)
Son of a bitch!   (clearly relieved)
                              Very nice recovery, Neil.
                              Posse will get there shortly.

              CHASE PILOT (COMMS)
       Yeah, might take a while.

2 <u>**EXT. LANDING STRIP 35, ROGERS DRY LAKE BED, EAFB - LATER**</u>   2

An Air Force jeep **WIPES FRAME, REVEALING** a FIRE TRUCK, a sedan
and TECHS surrounding Neil's X-15. AIR FORCE HELOS circle.

                    *Rogers Dry Lake
              Edwards Air Force Base, 1961

The period cars are cool, <u>but our eye is drawn to the X-15.</u>
<u>The long fuselage, the thick dorsal fin, the NASA signage...</u>
<u>It's every bit as awesome as the flight we just witnessed.</u>

PUSH IN on <u>THE OPEN COCKPIT</u>. NASA engineers take readings and
Neil makes notes, UNFAZED by the flight. As he gets out, **JOE
WALKER**, 40, thoughtful, walks up.

> JOE
> You okay?

> NEIL
> Yeah.

Joe looks at him, probing. But before he can say anything --

> FLIGHT SURGEON
> I gotta do his work up, Joe.

Joe nods. Neil follows the surgeon off. Joe turns back to
his jeep, only to find a grizzled AIR FORCE COLONEL, late 30s.

> ***** COLONEL YEAGER (O.C.)
> Kid's a good engineer, but he's
> distracted.

> JOE
> He got home, Chuck. He bounced off
> the atmosphere and still figured out
> how to get home.

The Colonel frowns. His tag GLINTS in the sun. **C. YEAGER.**

> COLONEL YEAGER
> Third mishap this month, Bikle
> should ground him before he hurts
> himself.

Yeager heads out. **HOLD ON** Joe, wondering if Yeager's right as
we **PRELAP** an **ODD WHINE**...

3   <u>**INT. DANIEL FREEMAN MEMORIAL HOSPITAL - INGLEWOOD, CA - DAY**</u>   3

CLOSE ON a 2-year-old girl (**KAREN**). Quiet, self-contained.
PULL BACK to see she's on a gurney, a **COBALT RADIATION MACHINE**
<u>HULKING</u> over her -- the source of the **MECHANICAL WHINE.**

HOLD for an AWFUL BEAT... then **REVERSE TO** the OBSERVATION
WINDOW. Neil and his wife, **JANET**, 27. <u>Watching.</u>

> NURSE
> We've found this to be effective in
> treating tumors like Karen's.

JIM: Colonel Chuck Yeager was not present for this particular landing of the X-15, but he was at Edwards Air Force Base in April 1962 when this flight took place, having taken over as first commandant of the USAF Aerospace Research Pilot School.

JOSH: We couldn't resist putting two of America's flying legends together, especially as Yeager was one of Neil's prominent detractors.

JIM: That's true. In Yeager's autobiography and subsequent interviews, he expressed several harsh sentiments toward Neil.

JOSH: Yeager's comment here is based on a quote from his autobiography that you include in your book: "Neil was a pretty good engineer but he wasn't too good an airplane driver." In post, we decided to pull back on this comment to help convey what the real issue was.

JIM: Yeager's evaluation of Neil's piloting skills is way too harsh in my opinion, but there was a series of mishaps in his flying in the months after Karen's death. In fact, there's some pretty good evidence that Paul Bikle, head of the Flight Research Center, considered grounding Neil as a result. Bikle also didn't write a letter of support for Neil's astronaut application.

TOP-BOTTOM: The crew at the X-15 landing, including Colonel Yeager (Matthew Glave).

JOSH: Other than Neil's small flying mishaps, it seems clear no one at work would have had the slightest idea Neil was dealing with such tragedy at home. We introduce Karen in the middle of cobalt therapy to emphasize the horror. As I said earlier, showing not only what Neil was going through at the time but also his ability to endure it with grace and dignity felt key to understanding who Neil was.

JIM: This is all historically accurate. Having tried unsuccessfully to combat Karen's tumor with X-rays Karen's doctors turned to the one possible treatment remaining: cobalt. The invention of the cobalt machine in the 1950s had been a major breakthrough. Unfortunately, side effects were very unpleasant and damaging, especially for a two-year-old girl. After the second treatment, it was clear that the cobalt machine was too much for Karen.

JOSH: From our conversations with Janet, it seemed that she, like Neil, was incredibly strong in the worst of circumstances. Claire Foy tried to make that clear in her portrayal.

JIM: As Neil's brother Dean once said: "Janet is as strong as horseradish." I think Claire does a superb job of exhibiting that!

THIS PAGE: Ryan Gosling on set with Lucy Stafford (Karen Armstrong).

First Man   POST CONFORMED BLUE   9.

*PUSH THROUGH THE WINDOW** to Janet. FIGHTING her emotions. She
reaches for Neil. He takes her hand...

Gone is his enthusiasm from earlier, <u>not the RELENTLESSNESS</u>.

> NURSE
> We're just not seeing the response
> we'd like yet.

Neil **EYES** the NURSE, setting the RADIATION DIALS, then looks
back to Karen.

HOLD ON Neil as the WHINE **CRESCENDOS.** Over the whine, **PRELAP**
the sound of **VOMITING** --

4   **INT. BEDROOM, ARMSTRONG CABIN, JUNIPER HILLS, CA - DAY**   4

Neil holds Karen as she vomits into a bucket. They're on her
bed, one of two kids' beds in the room. A curtain is all that
separates Neil and Janet's queen. Karen finishes, slumps.

> NEIL
> It's okay. It's okay. There you
> go, sweetheart.

Neil tries to soothe her, singing softly.

> NEIL
> *I see the Moon, the Moon sees me,*
> *Down through the leaves of the old oak tree.*
> *Please let the light that shines on me,*
> *Shine on the one I love.*

PUSH IN on Neil, on the PAIN in his eyes. CUT TO --

5   <u>OMITTED</u>   5

9   **INT. OFFICE, ARMSTRONG CABIN - LATER THAT NIGHT**   9

CLOSE ON a NOTEBOOK filled with TIDY NOTES. A mechanical
pencil writing out a heading. ***Cobalt Session No.2.***

> NEIL (INTO PHONE)
> Maybe I should talk to Dr. Johns...

FIND NEIL on the phone at a makeshift desk. The Aviation Week
to one side; Neil's now surrounded by neatly stacked **MEDICAL
BOOKS**, mimeographed RESEARCH PAPERS. <u>All marked and tabbed</u>.

>   JACK (OVER THE PHONE)
>   ...uh, who?

Neil grabs a medical paper on Cobalt therapy. Off the paper --

>   NEIL (INTO PHONE)
>   Harold Johns, he developed the
>   procedure. In Saskatchewan.

>   JACK (OVER THE PHONE)   NEIL (INTO PHONE)
> In Canada? How would you --   I could take time off work.

>   JACK (OVER THE PHONE)   NEIL (INTO PHONE)
> Maybe you should talk to the   I spoke to them already.
> folks at the hospital--

Neil pulls out his LOGBOOK from the hospital.

>   JACK (OVER THE PHONE)
>   Well. I'm sorry I couldn't be more
>   help.

>   NEIL (INTO PHONE)
>   That's okay, Jack. I appreciate it.

>   JACK (OVER THE PHONE)
>   Of course, Neil.

>   NEIL (INTO PHONE)
>   Give my love to June.

>   JACK (OVER THE PHONE)
>   You got it.

Neil hangs up. A beat. Then he turns back to his notebook.
We see him focus in on a column he's just written up...

**_Side Effects: Fatigue. Dizziness. Extreme headache. Vomiting
(repeated). Hair loss/Scalp irritation. Loss of appetite._**

Neil scans the list... then pulls a DOG-EARED RESEARCH PAPER
off the shelf. He reads, jots notes... *looking for a solve.*

7 **INT. FOYER/KITCHEN, ARMSTRONG CABIN - DAY**   7

Neil leads Joe Walker in. Joe carries a casserole. Neil
pulls some beers out of the fridge.

JIM: Neil did a lot of research on Karen's illness. She had a glioma of the pons, a malignant tumor growing in her brain stem. And the "Jack" that Neil spoke to over the telephone was his sister June's husband, Dr. Jack Hoffman, who was in private practice in Wisconsin. Several times during Karen's illness Neil phoned Jack to talk about Karen's condition and treatment.

JOSH: I'm not sure Neil ever suggested calling Harold Johns. But Neil was a relentless problem solver. Given how much he loved Karen we assumed that's how he would approach her illness.

JIM: But there is one giant fiction here... the cabin was far too small for an office.

JOSH: It's true. Rick Armstrong visited this set and was impressed (they all really slept in the same bedroom), but at the script stage he'd told us Neil just had a desk in a corner of the family room. We wanted a place where Neil could go to be alone.

THIS PAGE: In June 1961, two-year-old Karen Armstrong was diagnosed with a malignant tumor. She was treated with cobalt therapy.

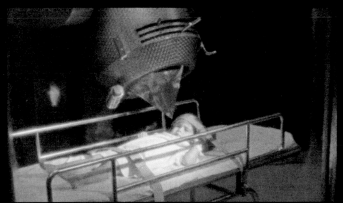

"PART OF DAMIEN'S VISION WAS TO GIVE AS MUCH AS A SENSE OF THE REAL PLACES THAT THIS STORY TOOK PLACE."

ADAM MERIMS, EXECUTIVE PRODUCER

ABOVE: Rehearsals were less about what was scripted and more about giving Ryan, Claire, Lucy, and Gavin a chance to be a family together. This footage wound up being used repeatedly in flashbacks in the film.

JIM: Joe Walker and Neil were close. In fact, the Walkers had lost their two-year-old son to a freak illness not long before. And Dick Day had been a flight simulation expert at the Flight Research Center and had worked closely with Neil before becoming assistant director of the Flight Crew Operations Division at the Manned Spacecraft Center.

JOSH: Neil wasn't sure he wanted to leave Edwards and the X-15 program.

JIM: No. In fact, Neil spent four or five months deciding whether to apply for astronaut selection. As Neil said, "It was a hard decision for me to make, to leave what I was doing, which I liked very much."

JOSH: But Dick Day and Walt Williams encouraged Neil to apply for Project Gemini.

JIM: Day did more than that. Neil credited Dick Day with getting him to transfer to Houston. And when Neil's application arrived after the deadline passed, Day (who was secretary of the Gemini Astronaut Selection Panel) slipped the paperwork into the pile with all the other applications.

First Man   POST CONFORMED BLUE   11.

                         NEIL
                    (off the casserole)
               Now who made that-- you or Grace?

                         JOE
               Grace did.

                         NEIL
               I'll take it then.   Thank you.

A small smile. Neil puts the casserole in the fridge, opens
the beers with a churchkey, hands one to Joe.

                       * JOE
               Dick Day called from Houston, he was
               asking after you.

                         NEIL
               Oh.   That about Gemini?

                         JOE
                    (nods)
               They're looking for pilots with a
               solid background in engineering.

Neil hesitates.

                         NEIL
               Well, maybe once Karen starts
               feeling better. It's just-- I
               wouldn't want to move her 'til then.

                         JOE WALKER
               Well. It'll be nice to keep you
               around.

If Joe has an agenda, Neil misses it. Off Neil --

8   **EXT. ARMSTRONG CABIN - DUSK**   8

Rick sits on the porch, playing with some tape.

PULL BACK TO FIND Janet standing nearby, smoking. We see the
TOLL Karen's illness has taken.

She takes another long drag as...

...Joe walks out. Janet puts on a good face.

JIM: Those who were at the cemetery recalled Neil being very tightly composed and stoic and showing little emotion. Janet, in contrast, was visibly shaken.

JOSH: Which is why we wanted to show Neil breaking down after the funeral. We wanted to show the human side of Neil, we wanted to get underneath the myth.

JIM: As Neil's sister June said to me, "It was a terrible time... When he can't control something, that's when you see the real person. I thought his heart would break."

JOSH: Stiff upper lip, that's how astronauts are portrayed and we know that's how they were in public. But in private, who could face this kind of loss without some emotion?

THIS PAGE: Karen Armstrong sadly passed away on January 28, 1962.

First Man   POST CONFORMED BLUE   12.

                    JANET
          Joe.

                    JOE WALKER
          Hey Jan. You, uh... hanging in?

                    JANET
          ...oh, you know.

He looks at her, nothing to say. Then --

                    JANET
          It's nice of you to come by.

He starts to go, when --

                    JOE WALKER
          Of course. Night now.

Joe gets into his truck. Off Janet --

6   **INT. BEDROOM, ARMSTRONG CABIN - MOMENTS LATER**   6

Neil sits beside Karen, sleeping at last. He strokes her hair
gently, staring down at her, glancing at a small name bracelet
she wears. Breathing into the respite.

Off Neil, we **PRELAP** the chilling sound of a **MECHANICAL CRANK** --

10   **EXT. JOSHUA MEMORIAL PARK - LANCASTER, CA - DAY**   10

CLOSE ON a **CRANK**. Turning. Lowering a small coffin.

TIGHT ON Neil with Janet and their son, Rick. All in black.
Janet cries as KAREN'S COFFIN sinks. Neil does not.

Neil keeps his eyes on the sinking coffin, HUGGING Janet
close. **TIGHT**. The SOUND of the crank takes us to --

11   **INT. LIVING ROOM, ARMSTRONG CABIN - DAY**   11

A LONG PUSH IN over trays of food. Dark suits and dresses.
Quiet murmurs. Some small talk.

First Man   POST CONFORMED BLUE   13.

Neil walks through, barely enduring it all. He moves to the
screen door, watches Rick toss a ball in the yard. Rick sees
his father, runs up to the door.

      RICK
Dad, wanna come play?

Neil hesitates. He looks at his son, <u>differently than before</u>.
A DISTANCE there now. A moment, then...

      NEIL
I should, I have to help your
mother.

Neil exits, escaping past Janet, <u>who's clocked it all</u>.

12 **INT. OFFICE, ARMSTRONG CABIN - SAME TIME**   12

Neil walks in, closes the door, closes the shade...

...and moves to the desk. His eyes **TICK** over the marked up
medical books and papers. His notebook, FILLED, lies open.

Neil CLOSES the notebook. Gathers the papers, STACKS THEM
AWAY. And then opens the DESK DRAWER. He pulls out NAME
BRACELET we saw Karen wearing earlier. And drops it in.

Then he **CLOSES THE DRAWER**. And sits. The **TEARS** coming,
quietly, so no one hears.

A beat, then as he pulls himself together, we **CUT TO --**

13 **OMITTED**   13

14 **EXT. BACK PORCH, ARMSTRONG CABIN - NIGHT**   14

Everyone's gone. Neil stares off. Janet walks outside. It's
COLD. She grabs a blanket, puts it around Neil, then sits. He
doesn't say anything, but <u>he puts an arm around her</u>.

Janet tucks into him. We linger on them, huddled together.

15 **INT. BEDROOM, ARMSTRONG CABIN - EARLY MORNING**   15

CLOSE ON a WESTCLOX ELECTRIC ALARM CLOCK, not quite six. **PAN
TO** Neil, staring up at the ceiling. <u>ALREADY DRESSED</u>. The
alarm TRILLS lightly. Neil sits up. Janet stirs.

     ✱NEIL
I thought I might go to work.

      JANET
Okay.

"IT WAS A TERRIBLE TIME... I THOUGHT HIS HEART WOULD BREAK."

JUNE ARMSTRONG, NEIL'S SISTER

JOSH: Karen was buried on Wednesday, January 31, 1962. While Neil did not return to work the very next day, he did go back to work the following Monday.

JIM: According to Grace Walker, it hurt Janet a great deal that Neil went right back to work: "She desperately needed her husband to help her. Neil kind of used work as an excuse. He got as far away from the emotional thing as he could. I know he hurt terribly over Karen. That was just his way of dealing with it."

THIS PAGE: Despite his grief, Armstrong returned stoically to work.

First Man   POST CONFORMED BLUE   14.

He exits. But we **HOLD ON** Janet, as she looks past the curtain to see Rick still sleeping. And the other bed -- **EMPTY**.

16 <u>**INT. TEST PILOT OFFICE, NASA FRC, EDWARDS AFB - EARLY MORNING**</u> 16

Neil sits at his desk, studying **AIRPLANE DIAGRAMS** for the <u>HANDLEY PAGE HP-115</u>. We note a CHANUTE AWARD on the desk.

> BUTCH (O.S.)
> They want a free flying trainer to
> simulate a landing? They haven't
> even figured out how to get there.
> I'm not wasting time on that.

Joe walks in with STAN 'BUTCH' BUTCHART, who flew the B-52 on Neil's X-15 mission. Butch clutches a **XEROXED MEMO**.

> NEIL
> Morning.

They look up at Neil, SURPRISED to see him.

> BUTCH
> Neil.

It's **AWKWARD**. Butch drops the memo, moves off. Joe lingers.

> JOE
> You can take a few days, you know.

> NEIL
> I know, I'm just getting up to speed
> on the new Delta wing in the UK.

Joe glances at the diagrams Neil's studying. Then, GENTLY --

> *JOE
> Bikle cancelled the trip. He wants
> you focused on writing up the pilot
> report from your last flight.

Neil looks at Joe.

> *NEIL
> Am I grounded, Joe?

> JOE
> (hesitates)
> Write up the report on the bounce,
> okay?

Joe moves off... but we **HOLD ON** Neil, watching Joe head into the corner office to chat with <u>FRC HEAD PAUL BIKLE</u>.

JOSH: Neil was indeed scheduled to fly the new Handley Page HP.115 delta wing in May 1962, but Paul Bikle, the FRC head, canceled Neil's trip.

JIM: Bikle cited Neil's workload, but, in truth, he was concerned about Neil's flying mishaps in the aftermath of Karen's death and didn't want Neil flying the RAF experimental jet. In fact, a number of Neil's colleagues thought that Bikle temporarily grounded him in late April 1962.

JOSH: Neil himself never believed Bikle grounded him. But we have Neil ask if he's grounded because we thought it was the best way to highlight what, according to your research, was going on with Neil professionally.

JIM: As I have already mentioned, Bikle also refused to support Neil's astronaut application. And Chris Kraft told me that, in retrospect, Joe Walker should have been aware of what was going on with Neil and grounded him. So, it's a bit surprising that Neil wasn't more concerned about his flight status.

JOSH: One final point here—Neil did begin working on the Lunar Landing Research Vehicle (LLRV) while at Edwards. We highlight that here because in many ways this was the true beginning of his work on Apollo.

BELOW: Joe Walker (Brian d'Arcy James) confers with Neil at his desk.

"IT'S INTERESTING TO SET THE STORY IN A DECADE WHICH DID SCARE THE HELL OUT OF YOU. AT THE TIME, WE ALL AT NASA WERE GOING, 'REALLY? YOU CAN DO THAT?' BUT EVERYBODY SAYS, 'OKAY, LET'S GIVE IT A SHOT. LET'S GO.'"

FRANK HUGHES, FORMER NASA CHIEF OF SPACE FLIGHT TRAINING

First Man   POST CONFORMED BLUE   15.

Neil **STARES**. To be stuck on the ground now... that's the last thing he wants. He **STRUGGLES**, staring down at his desk...

...noticing a **NASA X-PRESS NEWSLETTER** under a pile of papers.

CLOSE ON the newsletter, a headline clear: _**NASA TO SELECT ASTRONAUTS FOR PROJECT GEMINI**_

REVERSE TO Neil. Something **SHIFTING** in him. As he processes, considering the **CHALLENGE**, we **CUT TO --**

17   **OMITTED**   17

18   **INT. HALLWAY, ELLINGTON AIR FORCE BASE - LATER**   18

A hall lined with chairs. Candidates sit reading PROJECT GEMINI BRIEFING BOOKS. Most of them in **MILITARY UNIFORM**.

FIND NEIL, in a suit and tie, sitting apart from the others.

**Astronaut Selection, Project Gemini
Ellington Air Force Base
August 13, 1962**

ELLIOT SEE, 35, another suit and tie, walks in. Elliot's a cerebral flight test engineer from UCLA; he catches a few looks from the military men and decides to sit next to Neil.

> ELLIOT
> Civilian?

> NEIL
> Yeah.

> ELLIOT
> Yeah, me too. Elliot.

> NEIL
> Neil.

Neil turns to his packet, but Elliot, anxious, keeps talking.

> ELLIOT
> Tough morning, huh? I barely lasted
> two minutes in the ice bath. Course
> I suppose NASA's probably interested
> more in psychological reactions.

> NEIL
> Well, I think I made it pretty clear
> that I thought it was cold.

This draws a smile from Elliot. A **SMALL CONNECTION**.

First Man　　　　　POST CONFORMED BLUE　　　　16.

> DEKE
> Armstrong.

Neil looks up, sees **DEKE SLAYTON**, 37, MACHO back when that was appealing, at the end of the hall. Neil stands.

> ELLIOT
> Good luck.

> NEIL
> Thank you.

Neil walks down the hall, passing a few of the military men.

> PETE CONRAD
> 'nother egghead.

COLONEL **ED WHITE**, 32, a lanky Texan with easy-going charm, SMIRKS at NAVY CAPTAIN **PETE CONRAD**, 32, a wicked witted WASP.

> ED
> It's Easter.

Pete laughs. Neil ignores it, keeps going as we **CUT TO --**

19　**INT. CONFERENCE ROOM, ELLINGTON AIR FORCE BASE - MOMENTS LATER** 19

Neil sits in front of **BOB GILRUTH**, 48, bald, strong and calm and A PANEL OF MEN including Deke; CHRIS KRAFT, 37, clean cut; and **JOHN GLENN**, 40, very much the American hero and icon.

> GILRUTH
> Neil, we've been chatting with
> candidates about the program. As
> you know, our decision to forego
> Direct Ascent in favor of a Lunar-
> Orbit Rendezvous approach to the
> eventual Moon mission has had a
> major impact on Gemini.

> KRAFT
> Do you have any thoughts on that
> decision?

> NEIL
> Well, even considering Von Braun's
> initial criticism, it seems that the
> payload saved by parking the primary
> vehicle in orbit and sending a
> smaller ship down to the lunar
> surface is well worth the resulting
> risks and challenges.

JOSH: As Elliot mentions, NASA actually did have Gemini applicants hold a foot in an ice bath as long as was tolerable. We shot Ryan doing this, but cut the scene for pace. Pace is one of the hardest things to gauge at the script stage, in part because you can't predict what the actors will bring to the scene.

JIM: When Elliot See asks Neil if he was a civilian, Neil nods. Accurately. But, Neil had flown seventy-eight combat missions as a navy pilot in the Korean War. I think he would have mentioned that he had flown in the navy, although probably not the combat missions, he was way too modest for that.

JOSH: That's fair, but we wanted to highlight the fact that he and Elliot were the only civilians in the second astronaut group. We wanted to focus on their friendship.

JIM: Elliot, who was based in LA, did drive Neil down to Houston when they began work on Project Gemini. And they became very close working as the backup crew for Gemini V.

JOSH: We also introduce Ed White and Pete Conrad here.

JIM: Pete with characteristic aplomb!

JOSH: Of course, Ed was Neil's next-door neighbor and would become a good friend, despite being cut from a slightly different mold.

TOP LEFT AND RIGHT: Pete Conrad (Ethan Embry) and Edward Higgins White (Jason Clarke). Deke Slayton (Kyle Chandler) and Bob Gilruth (Ciarán Hinds).

                         KRAFT
              What do you see as the challenges?

                         NEIL
              Rendezvous and docking.

                         DEKE
              Why do you think spaceflight's
              important?

Neil pauses.

                       * NEIL
              I had a few opportunities in the X-
              15 to observe the atmosphere. And
              it's so thin; such a small part of
              the earth you could barely see it at
              all. When you're down here in the
              crowd and you look up, it seems
              pretty big and you don't think about
              it too much, but when you get a
              different vantage point, it changes
              your perspective... I don't know
              what space exploration will uncover,
              but I don't think it will be
              exploration just for the sake of
              exploration. I think it will be
              more the fact that it allows us to
              see things that maybe we should have
              seen a long ago but just haven't
              been able to until now.

It's a lovely sentiment. Everyone's impressed.

                         GILRUTH
              Does anyone have anything else?

                       *JOHN GLENN
              Yeah. Neil, I was sorry to hear
              about your daughter.

Neil nods. An awkward beat.

                         NEIL
              I'm sorry, is there a question?

                         JOHN GLENN
              Uh, what I mean is, do you think it
              will have an effect.

Neil pauses, considering. Then, genuine --

JIM: Deke Slayton, Walt Williams, and Dick Day were on the NASA selection panel for the second group of astronauts. John Glenn and Chris Kraft were not officially on the panel, but they did drop in on some of the interviews.

JOSH: We added Gilruth to help build his character, as he plays a large role in the story. Given that he was director of the Manned Spacecraft Center, it didn't seem like a stretch.

JIM: There's no record of this interview but in my conversations with him, Neil remembered only talking about the general nature of the space program.

JOSH: That's why we did a few introductory questions then moved to a general conversation about the Moon mission. But I can't take credit for most of this—a lot of what Neil says is taken from various historic interviews.

JIM: Neil was clearly deep. He was quite thoughtful about this journey and his role in it. You had a lot to work from.

JOSH: Amazingly, with all the research I did in the first two years I worked on the script, it was Ryan who turned me on to a bunch of these quotes—he'd found them while researching the part. I love when Neil says space exploration won't be an end in itself. That's a direct quote and it's such a poetic and powerful argument for space travel.

JIM: You also use this as an opportunity to highlight Neil's dry sense of humor.

JOSH: It helped lighten some of the exposition. But we wanted to take people back to the time when it wasn't obvious how (or even if) we'd get to the Moon, so we thought the exposition was necessary.

JIM: I will say, the interchange about Neil's daughter strains credulity a bit. Most of the panel members didn't know Neil had a daughter, let alone that she'd died. Dick Day, who did know, told me, "I never gave a thought to his personal affairs."

JOSH: This is one of those places fact is too strange for fiction. We're so focused on Karen that if the men on the panel didn't bring her up, I think they'd look negligent, stupid even. The audience wouldn't believe it.

That said, the panel doesn't dwell on it; not like I believe they would today. And it doesn't stop the panel from selecting Neil. This felt like a good opportunity to have everyone acknowledge the issues Neil's been having, and to demonstrate that Neil's such a strong candidate that the panel is going to look past those issues.

BELOW: In 1962 Armstrong applied for Project Gemini, NASA's second human spaceflight program.

First Man   POST CONFORMED BLUE   18.

> NEIL
> I think it would be unreasonable to
> assume that it would have no effect.

Off Deke, taking this in, **CUT TO --**

20   **INT. ARMSTRONG CABIN - NIGHT**   20

Neil, Jan and Rick eat supper. Quiet, a **PALLOR** still hangs.
The phone RINGS. Janet reaches for it.

> JANET (INTO PHONE)
> Hello? Yeah, sure. Neil?

Neil looks up. Janet holds out the phone. He takes it.

> NEIL (INTO PHONE)   RICK
> Yello.   (to Janet)
>   Can I go play outside?

Janet nods, motions for him to go. Rick scurries off.

> NEIL (INTO PHONE)
> Yes. Uh huh. Okay. Yes, sir.
> Thank you.

Neil hangs up. Processing. Then he sees Janet STARING.

> NEIL
> I got it.

He looks to Janet.

> JANET
> It's a fresh start.

> NEIL
> Are you sure?

> *JANET
> It'll be an adventure.

Janet **REACHES FOR HIM**. Neil holds her hand tightly, <u>wanting
to believe her</u>. Off this slightest touch of **HOPE**, we --

> **FADE TO BLACK.**

OVER BLACK

**Manned Space Center
Houston - Fall 1962**

JIM: On September 13, 1962, Deke Slayton called Neil and offered him a job as an astronaut in the second astronaut group, the "New Nine."

JOSH: In this scene, we're clearly leaning into the fact that Neil was (at least initially) unsure about leaving FRC to join Gemini. And the relationship between the death of Karen and that choice.

JIM: While Neil himself didn't remember any correlation, he did admit it was "a trying time." And Neil's sister June seemed to see a relationship. As she told me, "The death of his little girl caused him to invest those energies into something very positive, and that's when he started into the space program."

ABOVE: NASA's Astronaut Group 2, also known as the "New Nine," was the second group of astronauts selected by NASA and announced on September 17, 1962. Back row: See, McDivitt, Lovell, White, Stafford. Front row: Conrad, Borman, Armstrong, Young.

JIM: Bob Gilruth was a consummate engineer, I'm not sure he would have made such specific remarks about "Kremlin confidence." I think his references to the geopolitics of the Space Race would have been more general.

JOSH: But as the director of the MRC he felt like the right guy to provide context for the mission.

JIM: Sure. It's important to make the movie audience aware of the larger context of the U.S. space program. This sets the stage for Deke, who accurately articulates the basis for the Moon mission—picking a job so far out that the Russians, even with their "large rocket engines" would have to start from scratch.

JOSH: Another key here was for Deke to take us back to a time when the keys to Apollo—rendezvous, docking, and EVA (which are now common in space and space movies)—were anything but routine.

JIM: They were still figuring out if (and how) they could achieve those goals.

THIS PAGE: Gilruth shows the New Nine a NASA film. This scene is cut from the movie. A number of other cuts/omits were made over the next ten pages of the script. These scenes shot well, but were cut in post to drive the pace forward.

First Man   POST CONFORMED BLUE   19.

21   **INT. CONFERENCE ROOM, MSC – HOUSTON – DAY**   21

GRAINY FOOTAGE of the Russian cosmonaut program. Two Vostok capsules blasting off, one after the next. Above and below the footage, the words **SECRET/NOFORN**.

PULL BACK to **FIND** NEIL in the flickering light, taking **COPIOUS NOTES**; Gilruth's at front, a tech at a REEL-TO-REEL PROJECTOR.

> GILRUTH
> Two weeks ago, the Soviets put two manned spacecraft in orbit. At one point, Vostok 3 and 4 orbited within 75 miles of each other.

Footage of the two Vostoks. Elliot See, taking notes by Neil, looks GRIM, but Neil's FASCINATED. **GUS GRISSOM**, 39, gruff, flicks on the lights. We see Deke beside him.

> GILRUTH
> They're close on rendezvous. They can put nuclear payloads right above us every hour and a half.
> And space has given the Kremlin confidence. Support of the guerillas in Vietnam; what we now believe is a very real attempt to introduce missiles into Cuba...

Deke moves to the long blackboard at front, draws a circle way on the left and writes EARTH. He marks dots beside it for --

> *DEKE
> Sputnik 1, Sputnik 2, Vostok, Gagarin. The Soviets have beaten us at every single major space accomplishment. Our program couldn't compete, so we've chosen to focus on a job so difficult, requiring so many technological developments, that the Russians are gonna have to start from scratch.
> As will we. So instead of here...
>   (points to dots around Earth)
> ...we go here.

And now he walks all the way to the right edge of the blackboard... then past it to a **SEPARATE BLACKBOARD**... and draws a small circle. Which he labels **MOON**. He eyes the men.

> DEKE
> That's to scale. Check it.

A tech moves to the blackboard, measures as Deke continues.

First Man   POST CONFORMED BLUE   20.

> DEKE
> As you know, we've decided it's too
> hard to land one big rocket on the
> Moon. We're gonna need to put a
> smaller vehicle on the surface and
> get it back again, which means that
> we need men trained to catch the
> command ship and dock. That's the
> primary mission of Project Gemini.

Deke writes **RENDEZVOUS. DOCKING.** Neil, INTENT, takes it down,
more of that **FIRE** in him -- like when he saw the LLTV diagram.

> DEKE
> When we think you're ready, each of
> you will be assigned a flight with a
> specific task. Only after we master
> these tasks do we move on to Apollo
> and consider trying to land on the
> Moon.
>   (then)
> Gus, you got anything you wanna add?

Deke looks over at Gus. Gus shakes his head.

> GUS
> Just do your job.

Neil and THE NINE GEMINI MEN (Elliot, Ed White, Pete Conrad,
rugged former Navy test pilot **JIM LOVELL** and others) eye the
blackboard as the tech ERASES Deke's Moon and draws a new one.

> DEKE
> Almost to scale.

<u>A foot further away.</u> Off Neil, taking in the vast blackness
between earth and Moon, we **PRELAP** --

> WALTER CRONKITE (**PRELAP**)
> *Good evening, this is Walter
> Cronkite at CBS News Headquarters in
> New York. At its beginning this day
> looked as though it might be one of
> armed conflict...*

22   **OMITTED**   22

JOSH: So there are a few small fictionalizations inherent in these pages. For starters, in the script Ed and Pat White's house is across the street from the Armstrong's, whereas in reality the two houses were side-by-side.

JIM: Yes, the Whites and Armstrongs both arrived in Houston in the fall of 1962. After living in the Tara Apartments across from Hobby Airport for a year (Mark was born during this time in April 1963), the two families bought property together in the El Lago development. The two families actually bought three contiguous lots, and split the middle one in half so that each family had a lot and a half to build a house on.

JOSH: We found a great neighborhood in Atlanta that looked a lot like El Lago, but we couldn't find two adjacent homes that were suitable. We did find a home that would pass for the White's across from a vacant lot—so we built a home to match the Armstrong's on the lot.

JIM: Rick Armstrong was pretty impressed with the place. That big fireplace and the living room is pretty spot on.

JOSH: We felt comfortable consolidating the timeline and skipping over the period when the two families were renting property. While the specifics of how they met and came to live near each other were different, the through-line is the same— these were two astronaut families that became very close in the mid-60s during Project Gemini.

We've kept these scenes as originally scripted, but we've been playing with intercutting Scene 21 with Scenes 23 and 24, and then losing Scene 25 entirely. This is something we've tried for pace—we think it's liable to achieve the same result as the scripted version. We've left the script intact so readers can judge for themselves.

TOP: Pat White (Olivia Hamilton) welcomes Janet to the neighborhood.
MIDDLE: Production designer Nathan Crowley and his team built a replica version of the Armstrongs' El Lago house.
BOTTOM: Ed White (Jason Clarke) tosses a football on his front lawn with Eddie Jr. (Braydyn Helms).

23   **INT. LIVING ROOM, ARMSTRONG HOUSE – EL LAGO, TX – DAY**   23

CLOSE ON a B&W <u>CBS NEWS BROADCAST</u>. WALTER CRONKITE sits at his desk.

> WALTER CRONKITE (ON TV)
> *...between Soviet vessels and American warships on the sea lanes leading to Cuba. But there has been no confrontation as far as we know, and some hope has been generated by suggestions of negotiation.*

FIND NEIL watching, a Gemini binder open in front of him, a number of equations written in the margins.

Nearby Janet, recently pregnant, unpacks BOXES. Or rather, she was. Now, she looks from the TV to Rick, who's painting a candy bag for Halloween, a GUMBY DOLL on the floor nearby.

There's a **KNOCK** at the door. Janet walks over --

24   **INT./EXT. ARMSTRONG HOUSE – DAY**   24

Janet opens the front door. MEET **PAT WHITE**, 30, perky, blond, youthful, upbeat and warm.

> * PAT
> Hey. I'm Pat. Got here about a week before you, so welcome to the neighborhood...

Pat hands her a plate of cookies.

> JANET
> Oh, that's so kind of you. I'm Janet.

> PAT
> Nice to meet you.

Janet looks past Pat, spots a MAN tossing a ball with his son. <u>We recognize Ed White</u>. Janet watches for a moment, maybe a tinge of sadness at how far away that kind of normalcy seems.

> JANET
> (pointing)
> So that's you over there?

> PAT
> That's us right there.
> (pointing)
> That's my husband, Ed...
> (MORE)

[handwritten: INTERCUT WITH SC. 21]

                    PAT (CONT'D)
          and that's Eddie junior... I don't
          know where Bonnie is.

                    JANET
          You've got two.

                    PAT
          I do. I see...
            (noticing Janet's stomach)
          Is this your first one?

                    JANET
            (indicating Rick)
          Oh no, we've got a boy, Rick.

                    PAT
          How old is he?

                    JANET
          He's five and a half.

                    PAT
          We should get them together.

                    JANET
          Yeah, that'd be great.

A smile. It almost sets Janet at ease.

                    PAT
          Actually, I just got the kitchen
          squared if you want to join us for
          dinner?

Janet hesitates. She'd really like to. But she declines.

                    JANET
          Oh, that's real nice of you but I've
          already got some soup on, so...

                    PAT
          Okay. Well, another time.

                    JANET
          Yeah, I'd love that.

Pat smiles again. Janet smiles back.

                    PAT
          It was nice meeting you.

                    JANET
          And you, Pat. Bye.

25   **INT. LIVING ROOM, ARMSTRONG HOUSE - DAY**   25

Janet walks back in. Neil looks up.

> JANET
> Well, we've been invited over for
> dinner.
> (off his look)
> I said we were busy.

> NEIL
> (genuinely grateful)
> ...Thank you.

> JANET
> You want a cookie?

She holds out the tray. He takes one.

> JANET
> You okay with me picking up a pizza?

> NEIL
> Uh huh.

Janet never had dinner on. Neil returns to work...

26   **INT. MULTI-AXIS TRAINER ROOM, MANNED SPACE CENTER (MSC) - DAY** 26

WIDE ON a **HUGE, GEODESIC STRUCTURE** in a hangar; COLORFUL PIPES
cocooning a **COCKPIT CHAIR** SUSPENDED IN THREE CONCENTRIC RINGS.

> GUS
> The Multi-Axis Trainer was designed
> to replicate roll coupling on three
> axes, the kind you might encounter
> in space.

Find the Gemini astronauts walking in with Gus, Deke and a SIM
SUPE. Gus continues to the group.

> GUS
> The challenge is to stabilize the
> machine before you pass out.
> (to Neil)
> First victim, Armstrong.

Neil nods. He walks up the steps and sits in the cockpit
chair. The SUPE straps him in then starts the machine...

The inner rings **SPIN**, tossing Neil in all directions. Slowly
at first... then **FASTER**.

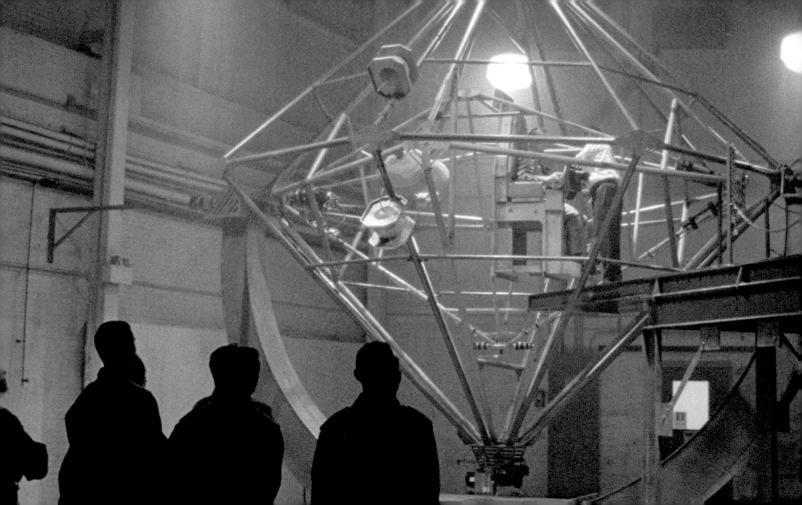

JOSH: We had quite the debate about using the Multi-Axis Trainer (MAT).

JIM: The MAT simulated the disorientation one would feel in a tumble spin during re-entry. But as none of the Mercury missions encountered this kind of spin, training with the MAT stopped being a core part of NASA's astronaut preparation after Mercury.

JOSH: So, it's unlikely Armstrong and his fellow Gemini astronauts trained on it.

JIM: Yes, but given the culture of the astronaut corps, the veterans might've wanted the new guys to experience the same "torture" they had. So, it's possible Gus or Deke unofficially arranged for the New Nine to give the MAT a spin. But it does beg the question—why not shoot a trainer we're sure the Gemini guys actually used?

"DAMIEN AND I CLICKED BECAUSE THERE'S A SORT OF GRITTINESS AND DIRTINESS AND REALITY... I WILL ALWAYS PUSH TO DO THINGS FOR REAL."

NATHAN CROWLEY, PRODUCTION DESIGNER

First Man   POST CONFORMED BLUE   24.

PUSH IN on Neil. The chair speeds up, his head WHIPPING in and out of frame as he struggles to analyze the spin.

His HANDS **GRAB** the controls... CLICKING to steady himself... but the machine just spins FASTER. Neil bores down... But it's <u>TOO FAST</u>... his eyes droop... and we **FADE TO GRAY.**

FLASH TO --

Juniper Hills. The cabin. Karen. And Janet, younger, carefree. Neil with them. With Karen. Happy.

A mechanical thud **SLAMS US BACK TO --**

THE MULTI-AXIS TRAINER. Neil blinks, **OPENS HIS EYES.** The Supe moves to unbuckle him. Deke looks to Ed.

> DEKE
> White, you're up.

> ED
> Yeah, I got it.

> NEIL
> I'm okay. Let's go again.

Neil STARES right at Deke. Gus eyes Deke. Who shrugs. Gus nods to the Supe, who tightens the straps, starts the machine.

As it spins, we **PUSH INTO <u>NEIL'S POV</u>.** The room, spinning. *And if it's nauseating, well, that's how it's supposed to feel.* Neil **CLICKS** the controls and the trainer slows...

...and then <u>SPINS</u> **FASTER.** Images of the other astronauts BLEED TOGETHER. As Neil starts to FADE, we **SMASH TO --**

27   **INT. BATHROOM, MSC - DAY**   27

Neil vomits. Head on the toilet seat. **SHAKING...**

Neil finally gets it under control. Pale, vulnerable. Almost overwhelmed. He spits into the toilet, then pulls himself up. Wavering for a moment, then staggering out to the sink...

...just as Ed walks in. Stares at Neil. Neil stares back, these two opposites facing each other. A beat.

Then Ed <u>turns GREEN</u> and **RUSHES INTO** a stall. We hear him **VOMITING.** Off Neil, maybe a small smile, **CUT TO --**

28   **INT. CLASSROOM, MSC - HOUSTON, TX - NIGHT**   28

CLOSE ON A THICK GEMINI BINDER, opening to --

JOSH: The Gemini guys put in days on the Johnsville Centrifuge—nicknamed "the wheel"—and in the "Vomit Comet"; weeks in desert and jungle survival schools; thousands of hours in simulators of all kinds. We wanted something people hadn't seen before, something that would get across both how unusual—and how grueling—this training was. Jim and I had a series of conversations with Damien, Frank Hughes, and others; in the end, we decided the MAT was the best way to get that across.

Plus, there was a fringe benefit. Neil and Dave Scott find themselves in exactly this type of spin on Gemini VIII, so this scene serves as great visual foreshadowing.

JIM: One other point—in the MAT they use these days at Space Camp they don't let you spin in one direction long enough for the fluid in your inner ear to shift, so you don't get nauseous.

JOSH: According to Frank Hughes that wasn't the case with the trainers NASA used. He said they'd occasionally spin continuously in one direction and wreak havoc on your inner ear, so there was a good bit of vomiting. Maybe not as much as in the Vomit Comet, but we show most of the guys succumbing to convey how physically challenging the training was.

TOP: A photo of the actual Multiple Axis Space Test Inertial Training Facility at the Lewis Research Center (now the John H. Glenn Research Center) in Cleveland.
BOTTOM: Technical Consultant Frank Hughes talks Ryan through the MAT training. Hughes started working for NASA in 1966 as a Simulation Supervisor or "Sim Supe"- with expertise in guidance computers. He eventually became Chief of Space Flight Training at NASA.

Physics of Rocket Propulsion - Rocket Vehicle Performance
1. Equations defining stage performance
2. Theoretical optimization of stages
3. Practical techniques using digital computers
4. Trajectory losses (drag, gravity, potential energy velocity)

PAN UP to Ed. Exhausted. Around him, the others scan binders with **BLEARY EYES**, many with **VOMIT STAINS** on their shirts.

Ed glances at Neil, behind him. Neil reads, **FASCINATED**. Just then, DAVID HAMMOCK enters, writes on the blackboard.

> HAMMOCK
> Gentlemen, welcome to basic rocket physics. We'll just be covering the first chapter tonight.

As Neil whips out a pen, intent, we **RACK BACK TO** Ed. STARING. And WONDERING: *Who the hell is this guy?* Ed shakes his head, flips open the chapter - 204 pages. Off Ed, **CUT TO** --

29   **OMITTED**   29

30   **INT. FOYER/HALLWAY, ARMSTRONG HOUSE - LATER THAT NIGHT**   30

Neil walks in, tired. He puts down his bag, sees the nursery door open. The outline of a crib. He hesitates, walks into --

A31   **INT. NURSERY, ARMSTRONG HOUSE - CONTINUOUS**   A31

Neil leans over the crib, stares down at the ten-month-old sleeping soundly. **MARK**. For a moment, Neil just stands over the infant. Frozen, a distance there. Off Neil, **CONFLICTED** --

31   **INT. KITCHEN, ARMSTRONG HOUSE - LATER**   31

Neil eats alone at the kitchen table, the house ASLEEP... save for Janet. She enters, sits with Neil.

> JANET
> Hi.

> NEIL
> Hi.

Neil nods to her, but he's DISTANT. Not oppressive, like after Karen died; but detached.

> JANET
> You okay?

> NEIL
> Yeah... Just thinking about this lecture... it's kinda neat.

"I THINK HE WAS MORE THOUGHTFUL THAN THE AVERAGE TEST PILOT. IF THE WORLD CAN BE DIVIDED INTO THINKERS AND DOERS—TEST PILOTS TEND TO BE DOERS AND NOT THINKERS—NEIL WOULD BE IN THE WORLD OF TEST PILOTS WAY OVER ON THE THINKER SIDE."

MIKE COLLINS

JOSH: We wanted to get across that a good part of the training took place in the classroom. Fortunately, NASA provided us with detailed schedules of Neil's training activities, including weekly Flight Crew Training Reports published by Raymond Zedekar, the Astronaut Training Activities Officer. This is how we know that David Hammock from the Spacecraft Research Division taught the Physics of Rocket Propulsion to the New Nine in their third week of training—on October 29, 1962, to be exact. One small fudge—it was actually taught from 8.30 to 10.30 on Monday morning, rather than at the end of a long day. Big thanks to Bert Ulrich, Bill Barry, and the rest of the folks at NASA for their help throughout the process.

JOSH: The theremin album *Music out of the Moon*, which opens with 'Lunar Rhapsody,' was one of two albums (recorded on cassette tapes) Neil took to the Moon with him.

JIM: *Life* magazine reported that Neil took it along for sentimental reasons, as he and Janet had allegedly fallen in love with the theremin music early in their marriage, when they lived in Juniper Hills. When I spoke with Janet and Neil, they both said they liked the music very much. While Neil said he only remembered bringing it along because it was fitting, Janet was fairly certain that Neil selected the music in part because no one else would have known it—so that when he played it, she'd know he was thinking of her.

JOSH: This was all a big surprise to us, but anyone who knows the Armstrongs knows they're quite musical. Both Mark and Rick play guitar and Neil's granddaughter, Kali Armstrong, is a fantastic singer. Needless to say, this was just one of those little things that we thought gave more insight into the man behind the myth.

JIM: While we don't know how much Neil talked to his colleagues about it, he was musical director of his fraternity and he did write and co-direct two musicals while at Purdue. The first of his shows, in fall 1953, was *Snow White and the Seven Dwarfs*, featuring music from the famous Walt Disney film. The second production was, indeed, *The Land of Egelloc*, alternately titled *La Fing Stock*, and featuring lyrics that Neil had written to the music of Gilbert and Sullivan.

> JANET
> What's neat about it?

> NEIL
> Well, it was about how to rendezvous
> with the Agena? If you thrust, it
> actually slows you down because it
> puts you in a higher orbit. So you
> have to reduce thrust and drop into
> a lower orbit in order to catch up.
> It's backwards from what they teach
> you as a pilot but if you work the
> math, it follows.
> (lost in it, a beat)
> It's kinda neat.

> JANET
> Yeah, it's kinda neat.

That LIGHT in his eyes, it's back. Janet tries not to <u>LAUGH</u>.
Neil catches it. Realizes...

> NEIL
> What's funny?

> JANET
> It's not funny. It's just... it's
> kinda neat.

She breaks; they both LAUGH. Off this, we **CUT TO --**

32 **<u>INT. LIVING ROOM, ARMSTRONG HOUSE - MOMENTS LATER</u>** 32

Neil stands at the record player. Puts on a record... We
hear an otherworldly melange of piano, chorale and THEREMIN.
As he stands, Janet walks out of the kitchen.

> NEIL
> Do you remember this?

> JANET
> Yeah. I'm surprised that you
> remember it.

He smiles. They stand a few feet apart, sizing each other up,
like kids at a school dance. Almost with fresh eyes.

He <u>holds out his hand</u>. She takes it and he <u>pulls her in</u>.
They SWAY together... it turns to **DANCING**. Neil's hands on
her back feel almost foreign...

...like they've forgotten what this is. And for a moment,
they're just a couple kids in love.

First Man   POST CONFORMED BLUE   27.

<u>He kisses her.</u> Sweetly, then with URGENCY. It's raw,
VISCERAL, real **HEAT** there. As they pull at each other, we
move to the phonograph, the spinning record: **"Lunar Rhapsody."**

33-42 **OMITTED (HOUSE FIRE SEQUENCE)**   33-42

A43   **INT./EXT. ED & PAT WHITE'S HOUSE - TWILIGHT**   A43

Pat and Janet prepare dinner as **MARILYN SEE**, 30s, teaches her
daughter (7) how to dance to _Peter, Paul and Mary_.

1965

Rick horses around with Ed's kids **EDDIE** (11) and **BONNIE** (8).
Elliot's other daughter (8) joins them as they run outside,
past Elliot, who smiles at them. Off Elliot, we **CUT TO** --

43   **INT. DINING ROOM, ED & PAT WHITE'S HOUSE - NIGHT** 43

The three couples eat, the kids still playing in the yard.

                    JANET
          Thank you for having us.

                    PAT
          It's a pleasure.

Carrie See runs out of the kitchen with a popsicle, trying to
keep up with Bonnie White.

                    MARILYN
          Carrie, slow down.

Carrie slows, runs her fingers along the banister, walks out.

                    JANET
          That's a lovely piano. Do you play?

                    PAT
           (shaking her head)
          Bonnie's taking lessons.

                    JANET
          Perhaps we'll sing for our supper.

                    ELLIOT
          Neil plays piano?

                    JANET
          Neil knows all sorts of show tunes.

               ED   NEIL
Come on.   Janet --

First Man   POST CONFORMED BLUE   28.

> JANET
> He was musical director of his
> fraternity in college. He wrote the
> musical for the all-student revue.

Neil's a bit embarrassed.

> NEIL
> I didn't write the music. We used
> music from Gilbert and Sullivan.

> *JANET
> He wrote all new lyrics. "The Land
> of Egelloc." It was quite funny.

> ELLIOT
> The Land of... Egelloc?

They look at Neil. A beat.

> NEIL
> Egelloc. You've never heard of it?

> ELLIOT
> I haven't.

> NEIL
> Oh, I'm surprised. It's a distant
> land, but it's a magical place...

> JANET
> It's college spelled backwards.

The whole group STARES at Neil.

> ED
> Seriously?

> NEIL (CONT'D)
> Yep.

The table laughs. Neil, too.

44   **OMITTED**   44

45   **EXT. BACK PORCH, ED & PAT WHITE'S HOUSE - NIGHT**   45

Elliot takes a **SEXTANT** READING. Ed and Neil sit behind him,
sharing a beer. The kids are still playing in the backyard.

> ED
> You're backup on five, huh?

"NEIL WAS A VERY SMART GUY. HE HAD VERY

First Man   POST CONFORMED BLUE   29.

>   ELLIOT
> Yep.

>   ED
> You'll get your own mission soon
> enough.

>   ELLIOT
> How's training going on Four?

>   ED
> Good. I think we're close on E.V.A.

>   ELLIOT
> First man to walk in space. That'd
> be something, huh?

>   ED
> Well, the walking's the easy part.
> It's getting back inside that's
> tough. Helluva ride if I come back
> with my tail hanging out.

>   NEIL
> Oh, I think McDivitt'll cut the cord
> before that happens.

A beat, then Elliot smiles. As does Neil.

>   ED
> Whoa, whoa, whoa, throttle back
> there, Armstrong.

The men laugh, a **BOND FORMING**. Pat comes to the door.

>   PAT
> Ed, phone for you.

>   ED
> Who is it?

>   PAT
> It's Deke.

Ed gets up.

"I THINK WE'RE GOING TO THE MOON BECAUSE IT'S IN THE NATURE OF THE HUMAN BEING TO FACE CHALLENGES. IT'S BY THE NATURE OF HIS DEEP INNER SOUL... WE'RE REQUIRED TO DO THESE THINGS JUST AS SALMON SWIM UPSTREAM."

NEIL ARMSTRONG

                         ED
          Okay.

Ed hustles into --

45A   **INT. KITCHEN, ED & PAT WHITE'S HOUSE - CONTINUOUS**   45A

Ed picks up the phone.   Janet's still at the table.

                         ED (INTO PHONE)
          Deke?   Ed.

45B   **EXT. BACK PORCH, ED & PAT WHITE'S HOUSE - SAME TIME**   45B

Neil takes a reading.   Elliot looks up.

                         ELLIOT SEE
          I think that's Castor, not Pollux.

                         NEIL
          I was testing you.   You passed.

                         ELLIOT SEE
            (smiles)
          Thought about letting it fly, but...

They laugh, sit.   Neil looks at the Moon... then hears Ed.

                         ED (O.C.)
          I'll call you back.

Neil looks towards the house.   Elliot turns as well... we see
someone has turned on the television in the kitchen.

We hear a **TELEVISION NEWS REPORT**...

                         CBS ANCHOR (V.O., ON TV)
          Leonov is tethered to Voshkod 2, but
          nothing separates the cosmonaut from
          space other than his pressure
          suit...

Neil and Elliot head inside.

46   **INT. KITCHEN, ED & PAT WHITE'S HOUSE - SAME TIME**   46

Ed stares at the TV, Pat, Marilyn and Janet watching with him.

                         CBS ANCHOR (V.O., ON TV)
          ...a pressure suit we're told was
          designed for the lunar surface.
          This is, of course, mankind's first
          E.V.A., or Extra-Vehicular Activity

JIM: NASA had wanted to make the first-ever spacewalk, or EVA (Extra Vehicular Activity) with Ed White on Gemini IV but cosmonaut Alexey Leonov and the Soviets beat them to it. (Incidentally, Neil considered "spacewalking" a "terrible term").

JOSH: Ed's EVA was a part of early drafts of the script, but we cut it to focus more squarely on Neil's journey. We have Ed talk about it here so we can see the fact that he was really one of the true astronaut pioneers.

JIM: Of course, while Ed was training for EVA in March 1965 no decision had been made on whether he'd actually attempt it on Gemini IV in June. Leonov's EVA was part of what pushed NASA to have Ed perform the EVA on Gemini IV.

JOSH: We make the newscast a bit more definitive to reinforce that Ed would indeed become the first American to walk in space. And to help explain his intense frustration at the Soviets beating the Americans to this particular milestone.

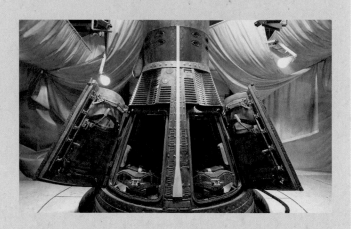

Neil and Elliot walk in, their smiles FADING as they see B&W FOOTAGE of **A COSMONAUT FLOATING OVER EARTH.**

The men TIGHTEN, watching the FIRST SPACEWALK, a huge feat. Like Gagarin and Sputnik, <u>a major victory for the Soviets.</u>

PUSH IN OMINOUSLY on the TV...

BANG! A fist **SLAMS** against the wall. Ed's fist. Janet and Marilyn **REACT**, startled.

> CBS ANCHOR (V.O., ON TV)
> *EVA is seen as one of the crucial tests the Astronauts must master if they are to successfully carry out their mission to the Moon. Astronaut Ed White was scheduled to perform the first EVA during Gemini 4, so this is yet another major victory for the Soviet Union in the Space Race...*

PUSH IN ON Neil and we CUT TO --

47   **OMITTED**   47

48   **INT. THE WHITE ROOM, PAD 19, CAPE KENNEDY - DAY**   48

CLOSE ON an Astronaut. Outside a Gemini cockpit. *Is Neil about to launch?*

GEMINI 5

REVEAL Neil and Elliot pulling themselves out of the cockpit, wearing <u>ROUTINE FLIGHT SUITS.</u>

They move away from the spacecraft, passing GORDO COOPER, 38, and Pete Conrad. In **SPACESUITS.**

> PETE CONRAD (COMMS)
> *We got it from here.*

Neil and Elliot walk out, towards the elevator. But Neil turns back, watching Gordo and Pete pull themselves into the Gemini spacecraft.

HOLD ON Neil looking at them, watching the techs close up the hatch. <u>Maybe wishing he were the one inside.</u>

First Man   POST CONFORMED BLUE   32.

49   **INT. LAUNCH PAD ELEVATOR, PAD 19 - MOMENTS LATER**   49

Neil and Elliot ride down the side of the small Titan rocket.

> ELLIOT
> Eight days up there... to be honest,
> I'm kinda glad we're backup on this
> mission.

> NEIL
> Hopefully they don't kill each
> other.

> ELLIOT
> It'd be quieter around here.

They trade a smile as the elevator stops...

50   **EXT. GEMINI LAUNCH PAD - SAME TIME**   50

Neil and Elliot emerge from the elevator and join the techs at
the base. We hear Deke call out.

> DEKE (O.C.)
>   Neil.

Elliot joins the techs, Neil walks over to...

ANGLE ON Deke with TWO MEN in their early 30s, **BUZZ ALDRIN** and
ROGER CHAFFEE. Neil joins them.

> DEKE
> Neil Armstrong, our backup
> Commander.

Neil nods, shakes hands. They introduce themselves.

> BUZZ   ROGER CHAFFEE
> Buzz Aldrin.   Roger Chaffee.

> DEKE
> A couple of the greenhorns from the
> third group. They'll be over in the
> blockhouse for launch. Neil, can I
> speak with you for a minute?
>   (to Buzz and Roger)
> Fellas.

Buzz and Roger leave. Neil follows Deke off to the side.

JOSH: This sequence highlights the fact that Neil wasn't among the first Gemini astronauts chosen to go to space, which we found interesting. It also highlights Neil's relationship with Elliot, introduces Buzz, and explains why Elliot wasn't assigned to Gemini VIII.

JIM: In reality, Neil and Elliot stayed to help Gordo and Conrad into the cockpit. Neil backed up Gordo, who got into the left seat. And Buzz was chosen as part of NASA's third group of astronauts in October 1963; this scene takes place sometime shortly before the launch of Gemini V on August 21, 1965, almost two years later. So Buzz and Neil would have met long before now.

JOSH: Buzz told me he was on roller skates in Ed White's driveway when they first met. I tried to work that in, but this was one of those times when compressing seven years into two and half hours requires skipping some events and conflating others. For example, putting a supporting character like Conrad closer to Neil even if he should be farther away so we can keep him alive. And, most importantly, keeping the focus on the story—which at this moment, is Neil and Elliot.

JIM: More specifically, the fact that Deke moved Elliot off Neil's team because he thought Elliot couldn't handle the EVA.

JOSH: Yes, given where we're going, we thought that detail was quite important.

>   DEKE
>   We're putting you in command of Gemini
>   8. Dave Scott is gonna be your pilot.
>   We get the Agena back on line, you're
>   probably gonna be the first to dock.

Neil nods, glances back.

>   NEIL
>   What about Elliot?

>   *DEKE
>   Don't worry about Elliot. We'll put
>   that brain of his to work, but we've
>   got a big EVA planned for 8. Dave's
>   a horse.

>   NEIL
>   Yes sir. Thank you.

Neil watches Elliot working with one of the techs. Off Neil, disappointed for him, we **CUT TO** --

A51   **OMITTED**   A51

51   **INT. KITCHEN, ARMSTRONG HOUSE - DAY** (2/28/66)   51

CLOSE ON hands washing dishes. We hear a Broadway show tune, maybe *Oklahoma!* It's something by Rodgers and Hammerstein...

REVEAL Janet washing, Neil drying. It rains hard outside, real Texas thunder. But in here it's lovely.

On the table, we notice a cake. *GOOD LUCK DAD!* An icing Gemini 8 approaches an icing Agena.

>   RICK
>   Dad, wanna come help?

REVEAL Rick (*now 9*) at the table, works on a PUZZLE of The Golden Gate Bridge. Mark (*3*) plays with trucks on the floor.

>   JANET   NEIL
>   Honey, let your father--   Sure.

Surprised, Janet watches Neil sit and start in on the puzzle. They work on it for a moment.

>   RICK
>   Have you ever been there?

Neil looks at Rick, who nods to the photo of the Golden Gate Bridge on the puzzle box.

First Man   POST CONFORMED BLUE   34.

       NEIL
  No. But I flew under it once.

Rick looks at him.

       RICK
Stop fooling.   NEIL
         I'm not.

      * RICK
  You can't fly under a bridge.

       NEIL
  Well, sure you can.

       RICK
  Mom, is dad making fun?

       JANET
  Doesn't sound like your dad.

There's a **KNOCK** at the door. Neil starts to stand but Janet,
wanting him and Rick to keep talking, heads for the door.

       JANET
  I'll get it.

FOLLOW JANET, glancing back at Rick and Neil. She can barely
hold back a smile. She opens the door. It's Ed, RAIN-SOAKED.

       JANET
  Hey, Ed.

       ED
  Oh, hi. Can I speak with Neil?

       JANET
  Yeah, sure. Why don't you come on
  in? You're soaked?

       ED
  No, I'll wait here. Thanks.

The mood shifts. Something's wrong. Janet knows not to ask.

> # "IT FEELS IMMEDIATE AND IT FEELS LIKE YOU'RE LIVING THROUGH THIS AS OUR CHARACTERS ARE LIVING THROUGH THIS."
>
> ISAAC KLAUSNER, EXECUTIVE PRODUCER

JOSH: We've jumped time here. Gemini V launched in August 1965 and Elliot See and Charlie Bassett crashed in February 1966. I'd originally written a Gemini VIII training sequence with Neil and Dave, but we decided early on it was redundant. We use the cake to try to convey the time jump—Neil was assigned to Gemini VIII in the last scene, now he's about to launch.

JIM: Neil flew not under the Golden Gate Bridge but under the Bay Bridge separating San Francisco and Oakland. He did it on his last navy flight on August 23, 1952, the day he left active service. It was a celebratory, unauthorized pass under the Bay Bridge's western span, which has a clearance of less than 220 feet.

JOSH: We loved this detail, which we found almost as surprising as Neil's musicality, but we couldn't find a period puzzle of the Bay Bridge. So, yes, we fudged it a little, but the idea is the same—a surprisingly daring move for a typically controlled individual.

JIM: It's also unlikely that Neil got the news about Elliot at home.

JOSH: That's right. Elliot crashed on the morning of February 28, 1966, less than three weeks before the launch of Gemini VIII.

JIM: Neil spent that morning in the Shuttle Mission Simulator (SMS) with Dave Scott. He was probably already hard at work.

JOSH: We wanted to see Neil get this news in an unguarded moment so we could show that loss truly weighed on him and Janet. And we had Ed bring him the news because we wanted to highlight the fact that one of the three men we just saw drinking beers together was suddenly gone. In prior treatments of the Space Race films, astronauts have shrugged off these kinds of tragedies. Which might have been the case in public. But we wanted to highlight how hard this must have been for them and their wives.

First Man   POST CONFORMED BLUE   35.

52   **EXT. FRONT STOOP, ARMSTRONG HOUSE – DAY**   52

Neil walks out onto the stoop.

> NEIL
> You know they make this thing called
> an umbrella, comes in real handy at
> times like these.

> ED
> Hey. I got some bad news about
> Elliot.

> NEIL
> I know, Deke told me he bumped
> Elliot, but he and Charlie are the
> crew on 9 --

> ED
> Neil. Elliot and Charlie were
> flying into St. Louis to train this
> morning. Their T-38 crashed on
> approach.

Neil stares, knowing from Ed's tone that Elliot's gone.

> ED
> There was a lot of fog...

Off Neil, processing...

53   **INT. KITCHEN, ARMSTRONG HOUSE – DAY**   53

Neil walks back in.

> RICK
> So did you really fly under that
> bridge? Was it fun? Were you
> scared?

Neil looks down at Rick. Lost. Janet sees something is
wrong, intercedes.

> JANET
> Ricky, honey, why don't you go get
> your homework so I can check it.

First Man   POST CONFORMED BLUE   36.

                   RICK
        Yes, ma'am.

Rick heads off. Janet glances at Mark, still playing with his trucks, then turns to Neil. Quietly.

                  JANET
        Who was it?

                  NEIL
        Charlie Bassett. And Elliot.

Janet PALES. **CUT TO --**

54   **INT. ELLIOT & MARILYN SEE'S HOUSE - TIMBER COVE - LATE AFTERNOON**54

CLOSE ON Marilyn See. **LOST.** **PULL BACK** to **FIND** a **PHOTO OF ELLIOT** and flowers on a mantle. Trays of food, **MOURNERS** in black, a priest. And a dazed old couple, Elliot's parents.

A few KIDS, oblivious, scurry through the crowd and out to --

55   **INT./EXT. PORCH, ELLIOT & MARILYN SEE'S HOUSE - LATE AFTERNOON** 55

The kids race into the backyard, passing Neil on the porch, looking through the screen door at the gathered mourners. He nurses a glass of CHIVAS REGAL.

               PETE CONRAD
        I was cornered by three congressmen
        at Arlington. They thought it was
        the time to ask why we don't send
        machines to the Moon instead? Two
        weeks before we launch 8.

Just beyond Neil, a CIRCLE OF ASTRONAUTS stand on the porch, passing a bottle of vodka, a lime inside.

               GRISSOM
        Shit, how the hell did this happen?

               JIM LOVELL
        Cernan told me the cloud cover was
        down to 500 feet. Probably never
        saw the building.

              * BUZZ
        Clearly, the error was the approach.
        He was coming in too slow to reach
        the runway.

Lovell glances at Buzz, taken aback. The men go quiet.

> BUZZ
> What? You know Deke had doubts
> about him. That's why he moved
> Elliot off Eight.

> NEIL
> Deke gave Elliot his own command.

They all look up. Surprised at Neil, who's normally taciturn.

> BUZZ
> Elliot wasn't aggressive enough.
> You of all people have to know that--

> NEIL
> No. I don't. I didn't investigate
> the crash, I didn't study the flight
> trajectory, and I wasn't the one
> flying the plane, so I wouldn't
> pretend to know anything.

> BUZZ
> We'll never be 100 percent sure.

Silence. The other men just sit there. The tension is thick.

56   **INT. LIVING ROOM, ELLIOT & MARILYN SEE'S HOUSE - DUSK**   56

Neil, even more on edge now, moves through the mourners...
eyes **TICKING** *past* **WIVES** *raising hands in greeting...*

...past **ED** *waving at Neil from across the room...*

...past **ELLIOT'S PHOTO** *on the mantle...*

...and pausing on **A LITTLE GIRL** playing jacks under a table.
Neil stares at the girl. She looks up at him. *It's Karen*.

Neil blinks at her, then quickly moves into --

57   **INT./EXT. KITCHEN, ELLIOT & MARILYN SEE'S HOUSE - DUSK**   57

Janet pours coffee for guests. Neil walks up beside her.

> NEIL
> Can we go?

> JANET
> Uh, not right now, I want to help
> Marilyn clear all this up, I don't
> want her to have to do it after
> we've gone.

JIM: On March 2, 1966, two weeks before the launch of Gemini VIII, Neil and Janet joined mourners at separate services for Elliot See and Charlie Bassett. While we don't know if remarks like this were made at his wake, certain astronauts definitely made negative remarks about Elliot's piloting abilities at some point.

JOSH: I got these comments from Deke Slayton's posthumously published autobiography *Deke! U.S. Manned Space, From Mercury to Shuttle*. I heard similar things from a number of astronauts, many of whom were close to Charlie Bassett and blamed Elliot for his death.

JIM: Neil had heard that others thought Elliot's piloting—particularly his instrument skills—were not as good as they should've been. But Neil flew with Elliot a lot and didn't recall anything of substantial concern. Neil grew especially disturbed when a couple astronauts seemed to blame Elliot for his own death. As Neil said, "It's easy to say what he should have done, but there might have been considerations that we

were not even aware of. I would not begin to say that his death proves the first thing about his qualifications as an astronaut."

JOSH: As a side note, Buzz was known for sharing the unvarnished truth (or his opinion of it). While other astronauts might have felt this was inappropriate there's a certain bravery in how Buzz shares his convictions. It's also a useful tool for us as storytellers—having a guy who's willing to tell the truth when others aren't. And, of course, Buzz's approach to social situations was pretty much the polar opposite of Neil's. In short—as we believe Buzz held this view of the crash—I used this moment to introduce that type of behavior, and to set up what was a natural tension between Neil and Buzz's interpersonal style.

> **"NEIL WAS PATIENT WITH PROCESSES. SOMETIMES HE COULD BE IMPATIENT WITH PEOPLE WHEN THEY DIDN'T MEET HIS STANDARDS."**
> MIKE COLLINS

TOP: Neil, Buzz (Corey Stoll), Jim Lovell (Pablo Schreiber), Gus Grissom (Shea Whigham), Roger Chaffee (Cory Michael Smith), John Young (Choppy Guillotte) and other NASA employees gather at Elliot's wake.

"IN THE PURSUIT OF EXCELLENCE, THERE'S ALWAYS A COST."

MARTY BOWEN, PRODUCER

JIM: Neil was close to Elliot and upset by his death, but Neil leaving the wake as he does, that's not directly based on research.

JOSH: No, it's a dramatization to emphasize how tough these losses were. The astronauts were stoic; you might not have seen their pain. But it's hard to believe they didn't feel it.

JIM: I thought it interesting that you decided to raise the specter of Karen again.

JOSH: Damien and I were both struck by the losses that Neil (and Janet) endured over his career as a test pilot and astronaut. For us, these seemed connected to a greater sense of loss, one we felt when we first read about Karen in your book.

JIM: It's true that the rate of fatalities among test pilots was extremely high during Neil's early flying career. In 1948 alone, thirteen test pilots were killed at Edwards. In 1952, sixty-two test pilots died in the US. In 1956, Neil personally witnessed Milburn G. Apt's fatal X-2 Starbuster crash and Iven C. Kincheloe's fatal F-104A Starfighter crash in 1958.

JOSH: We didn't want to gloss over those losses. On the contrary, we wanted the audience to feel how Neil (and Janet) must have felt in the wake of these tragedies. In fact, the Janet quote about Iven Kincheloe comes directly from our conversations with her.

TOP LEFT: At Elliot's wake, a photo of Elliot See (Patrick Fugit).
BOTTOM LEFT: A grave Bob Gilruth and Deke Slayton.
TOP RIGHT: Neil grieves alone by his pool in the backyard.

>   NEIL
>   (quiet)
>   I need to go.

>   JANET
>   Okay, well, why don't you go and sit
>   down and I'll bring you a cup of
>   coffee. It'll just be a minute.

She keeps pouring... until she realizes Neil's gone.

>   JANET
>   Neil?

She turns. We hear a car start. Janet looks out the window,
sees Neil pull away. Off Janet, we **CUT TO --**

58 **INT. PAT WHITE'S STATION WAGON (MOVING) - HOUSTON, TX - LATER** 58

Close on Janet. In the backseat. Upset and embarrassed.

>   JANET
>   I'm sorry. I hate to be a bother.

>   ED   PAT
>   Oh, Jan...   It's no bother.

Ed drives. Pat beside him. A beat.

>   *JANET
>   Neil's... there was a year when we
>   were at Edwards. Four pilots died.
>   (then)
>   We got good at funerals that year.
>   We haven't been to one in a while...

She looks out the window. A beat.

>   JANET
>   Has he ever talked to you about
>   Karen, Ed?

>   ED
>   ...Not really. No.

>   PAT
>   (a beat, gently)
>   Does he talk to you about her?

>   JANET
>   No. Never.

Off Janet, staring out the window --

59   **EXT. BACKYARD, ARMSTRONG HOUSE - LATER**   59

Neil stands under the stars.   He looks up at the Moon.   Then
starts practicing, taking readings with the sextant.

We see Janet in the window.   **HOLD ON** her.   She considers going
to Neil, but decides against, goes to help Rick with homework.

REVERSE BACK TO Neil.   Sextant raised.   Staring up at the
heavens.   Off Neil, we **PRELAP** --

                    PAO ANNOUNDER (**PRELAP**, LOUDSPEAKER)
               T minus one minute and counting on
               the Atlas-Agena launch...

A60   **OMITTED**   A60

B60   **OMITTED**   B60

60   **OMITTED**   60

61   **OMITTED**   61

62   **INT. WHITE ROOM, PAD 19, KENNEDY SPACE CENTER (KSC) - 10AM**   62

CLOSE ON the elevators.   They open and we **REVEAL** --

Neil.   **CLOSE ON** HIS FACE.   He looks straight ahead, STARING at
something.   A beat, then we **REVERSE TO** --

THE GEMINI 8 CAPSULE.   Hatch doors OPEN.   Backups in flight
suits doing final checks.   FAINT COMMS buzz.

                    Gemini 8
                    Two weeks later

This is no test.   **THIS IS IT.**

IN THE ELEVATOR, Neil grabs his pack.   Dave does the same.
Deke, beside them, gives Neil **A LOOK:** *You good to go?*   Neil
gives a subtle nod, then leaves Deke behind and...

...walks across the small bridge into THE WHITE ROOM.   He's in
helmet, spacesuit; Deke, **DAVE SCOTT**, 33, muscular and
handsome, and a number of techs behind him.

We feel a SHAKING and hear a distant **ROAR**.   Neil turns.
Through the canopy window, he sees a **FARAWAY ROCKET LAUNCH**.

                    AGENA CONTROL (COMMS)
               Liftoff.   Agena is go.
                    (then)
                         (MORE)

JOSH: The Gemini VIII mission was a huge challenge. While we had the Public Affairs Office (PAO) transcript and Technical (TEC) Air-to-Ground Ground-to-Air and On-Board transcript, these transcripts are far from perfect.

JIM: You had to check and double check the transcripts with the experts.

JOSH: We spoke to a number of astronauts and even reviewed certain portions with Rick and Mark Armstrong to try to verify who said what. But the transcripts only got us so far. The PAO transcript starts before launch, but there's no record of pre-launch comms in the TEC transcript. And Bill Barry in the NASA HQ historian's office couldn't find them, either.

JIM: There is a little bit of pre-launch comms in the video of the mission online.

JOSH: Yes, there are four hours of film from the Gemini VIII mission on YouTube, which include five minutes of pre-launch comms. We use some of that here. For the rest, Dave Scott put us in touch with Apollo Flight Director Gerry Griffin, who was in Mission Control for Gemini VIII.

Gerry was incredibly helpful both with pre-launch and Mission Control comms.

JIM: Neil and Dave were already in the Gemini cockpit when the Agena launched. And the launch probably didn't shake the White Room lights.

JOSH: We wanted to make it completely clear that this mission involved two launches within hours of each other, to highlight that achievement.

JIM: The parachute harness story is true. There was some epoxy in the catcher mechanism on Dave's harness and it wouldn't buckle. Even something as minor as that could have cost Neil and Dave the launch. Fortunately, after a little sweating, backup commander Pete Conrad and pad leader Guenter Wendt got the catch unglued by using a dental pick.

JOSH: This surprised us. Why would Wendt have had a dental pick? But several of our technical advisors confirmed that this is what he used.

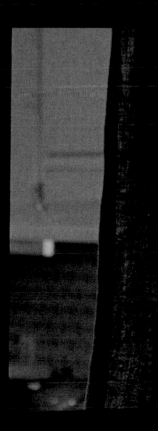

First Man   POST CONFORMED BLUE   40.

> AGENA CONTROL (COMMS)
> *Flight dynamics for Unmanned Agena*
> *Target Vehicle looking good, stand*
> *by for Gemini launch.*

Neil watches, then turns back and finds himself in front of the hatch door. A beat, then Neil moves forward and with the techs' help, pulls himself into the left hand seat of --

63   **INT. GEMINI 8 CAPSULE, PAD 19, KSC - CONTINUOUS**   63

Neil's feet are pointed up, he and Dave sit facing skyward, listening to the comms tracking the Agena as techs and backups hover over them, strapping them in.

> GUAYMAS CAPCOM (COMMS)
> AFD, Guaymas read you loud and
> clear. We have S Band track and...

Carnarvon capcom fades to STATIC

ASST FLIGHT DIRECTOR (COMMS)
Did you say all systems go on
T.M.?

                GUAYMAS CAPCOM (COMMS)
                ...we're having a little
                trouble locking up right now.

ASST FLIGHT DIRECTOR (COMMS)
Roger.

                PETE CONRAD (O.C.)
                Hold still, wouldja?

Neil looks over. Pete Conrad struggles to buckle Dave in. We see the catch on DAVE'S PARACHUTE HARNESS is **CLOGGED**.

RICHARD GORDON
What is that? Glue?

                PETE CONRAD
                Hold on a sec. Scoot down.

DAVE SCOTT
What are you doing?

                PETE CONRAD
                Hey, does anybody got a Swiss
                Army Knife?

> DAVE SCOTT (INTO COMMS)
> What'd you say? A Swiss Army Knife?

Pad leader **GUENTER WENDT** (42, spectacles, bow tie) leans in.

PETE CONRAD
Yeah, yeah. It's just a
little --

                GUENTER WENDT
                See if this'll do the trick.

He holds out a **DENTIST'S PICK**. Pete takes it. Dave stares as Pete grabs the harness catcher mechanism, digs out some epoxy.

First Man   POST CONFORMED BLUE   41.

                    DAVE SCOTT
Are you kidding me?   PETE CONRAD
                                      Got it.

        GUAYMAS CAPCOM (COMMS)
AFD, Guaymas Agena is go...   AFD (COMMS)
                                      Roger, Guaymas.

The men finish up, attaching hoses, then pulling back. Giving
everything a final once over before reaching for the doors.

We're ON NEIL as he looks across the small cabin and sees...
Dave's door **CLOSING SHUT** with a thud.

Neil turns, sees a tech above him nod. He gives a thumbs up.
Then his door **CLOSES IN** on him...

...and **THUDS SHUT.**

It's like being buried alive, worse when we hear the **SCRAPING
METAL** of the ratchet that **SEALS** the doors.

The capsule is now SEALED and **CLAUSTROPHOBIC.** Dave adjusts
his suit flow for air temp. Neil is still.

PUSH IN on Neil. On his eyes. Focused. DETERMINED. **HOLD
THERE** for a moment... then we hear a **LOUD MECHANICAL SOUND.**

Neil glances up; the GANTRY **SLOWLY PULLS** away from the rocket,
revealing CLEAR BLUE SKY. It's almost surreal... More so
when a **SEAGULL SQUAWKS,** hovering above Neil's window.

                    DAVE SCOTT (INTO COMMS, O.C.)
            Switching to HF. 1, 2, 3, 4, 5. 5,
            4, 3, 2, 1, check out.

Neil's eyes follow the seagull. A beat, then --

                    GT-8 STC (COMMS)
            *Copy. T minus 2 minutes. Engines
            to start.*

The focus **RETURNS** as Neil scans the console. The engine
lights come up.

        DAVE SCOTT (INTO COMMS)   EECOM (COMMS)
Ground power removal...   *Pressurization initiated.
                                    Ground power removed.*

Dave adjusts his suit flow again, Neil pulls out his **MIRROR.** A
beat. Neil hears a **BUZZING...**

...spots **A BUG** on the console. Mundane. Odd.

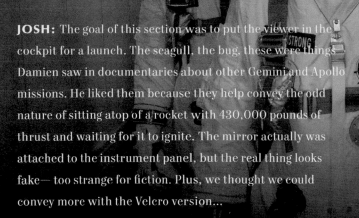

JOSH: The goal of this section was to put the viewer in the cockpit for a launch. The seagull, the bug, these were things Damien saw in documentaries about other Gemini and Apollo missions. He liked them because they help convey the odd nature of sitting atop of a rocket with 430,000 pounds of thrust and waiting for it to ignite. The mirror actually was attached to the instrument panel, but the real thing looks fake— too strange for fiction. Plus, we thought we could convey more with the Velcro version...

JIM: The comms here also came mostly from Gerry Griffin?

JOSH: Everything until liftoff. We couldn't have done it without him.

JOSH: Once we get to liftoff, things get easier on the script level, as the comms come directly from the Gemini VIII TEC transcript. Of course, it's not easy condensing an 8-hour flight into fifteen minutes, but the raw materials are there.

JIM: There are varying accounts as to the quality of the ride on the Gemini Titan. But most astronauts agreed that the acceleration right off the pad was pretty brisk.

JOSH: "Swift kick in the tail" is my favorite description. Which suggests more g-force than vibration. We do push vibration a bit, but only to help get across how intense these launches would seem from inside the cockpit.

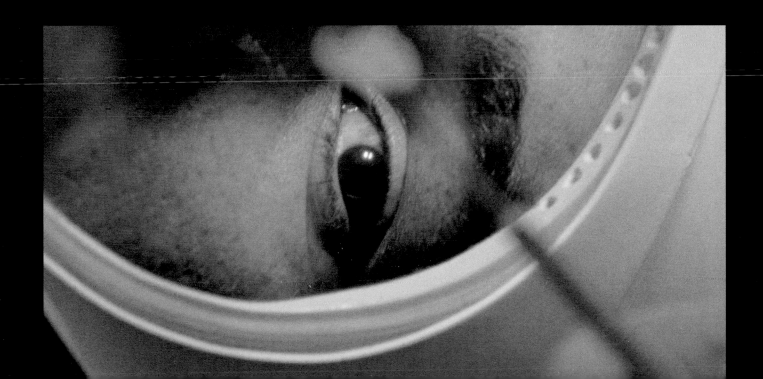

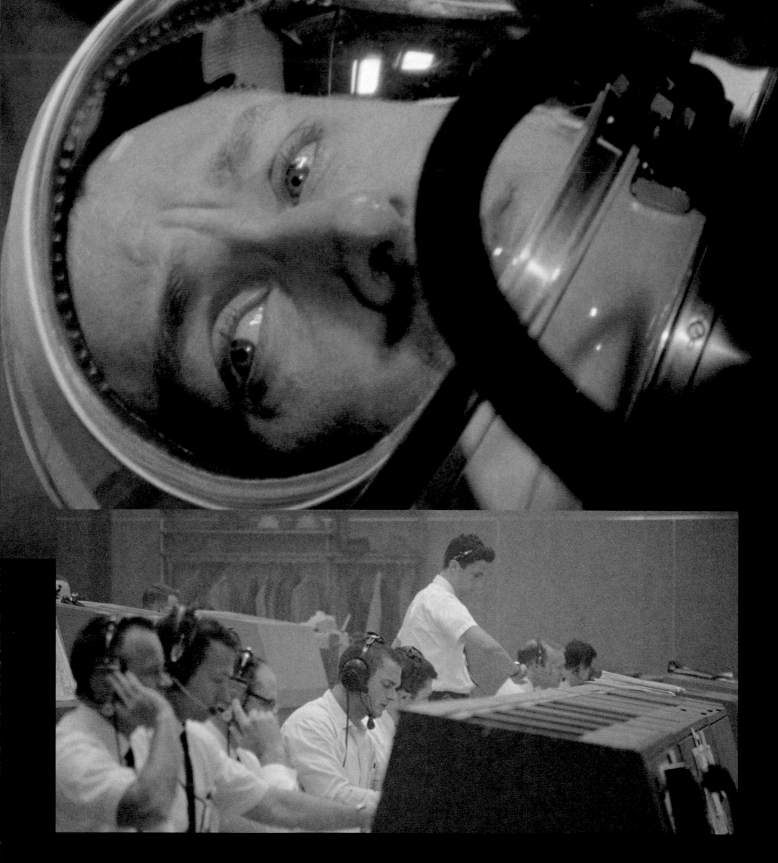

TOP: Neil watches the techs close the Gemini VIII hatches.

BOTTOM: Flight Director John Hodge (Ben Owen) oversees Mission Control.

First Man   POST CONFORMED BLUE   42.

                         LVTC (COMMS)
                *Stage 1 pre-valves coming open, 5
                seconds.* **T minus 20 seconds mark...**

Dave and Neil SET for launch. We feel the weight of the last
minutes, months, <u>years</u>...

                         LCC PAO (COMMS)
                *10, 9, 8, 7, 6... Main engines
                start...*

A **DULL THUNDER** from ten stories below turns to a **ROAR**...

                         LCC PAO (COMMS)
                4, 3, 2, 1... Ignition...

...and we feel a **JOLT** as the Titan **JERKS** off the launch pad.

                         LCC PAO (COMMS)
                Lift-off! *Lift-off 16:41:00!*

The **THRUST** kicks in, **SHOVING** them into their seats as they
vault away from the ground below.

                         NEIL (INTO COMMS)
                Clock is running. Got a Roll
                Program in.

                         LOVELL (CAPCOM, COMMS)
                Roger. Roll. Good liftoff, 8.

The **STRAIN** on Neil's face tells us *we're accelerating to*
<u>*18,000 MILES PER HOUR*</u>.

       NEIL (INTO COMMS)
Pitch program.   LOVELL (COMMS)
                                       Roger. Pitch program.

Neil switches mode as Dave checks the gauges, the computer.

    DAVE SCOTT (INTO COMMS)
DCS update received.   LOVELL (COMMS)
                                       Roger. DCS.

                         NEIL (INTO COMMS)
                Stage 2 tanks look good. That's
                about three and a half Gs.

Dave looks out the window, **STRUCK** by BLUE SKY <u>TURNING BLACK</u>.

       LOVELL (COMMS)   NEIL (INTO COMMS)
Go from the ground for   (enters staging command)
staging.   Roger.

A **SHEET OF FIRE ENGULFS** the craft... *IT'S FUCKING TERRIFYING.*
Dave FLINCHES... even Neil BLINKS. *What the hell is going on?*

A beat, then... as the fire outside subsides, we realize *this
is normal*. They refocus on the gauges.

> NEIL (INTO COMMS)
> Fuel cells are solid.

They hurtle forward, checking gauges, monitoring the stage...

> LOVELL (COMMS)
> *Gemini 8, you're go from the ground.
> Mark. V/VR = point zero eight.*

> NEIL (INTO COMMS)
> Okay. Mode 3.

The second stage cuts off; they're **TOSSED INTO MICROGRAVITY.**
Dave GRUNTS. Engines off, it's eerily QUIET. Dave pulls out
a mission checklist and it FLOATS across the cabin.

> NEIL (INTO COMMS)
> We've had SECO.

Neil looks out the window. *Nothing like the X-15, the world
much farther below.* The chaos down there, it's gone up here.

PUSH IN on Neil, A BOY, **STRUCK** by the **GRANDEUR**, the **MYSTERY**.
HOLD ON him for a moment, staring out the window.

A64   **EXT. GEMINI VIII - SAME TIME**   A64

We peer out over the nose of the craft, looking down at the
Earth. The view is breathtaking, utterly majestic.

B64   **EXT. FRONT PORCH, ARMSTRONG HOUSE - HOUSTON, TX - DAY**   B64

Rick stands below the flagpole, raises an American Flag. He
looks up at it, waving in the breeze.

C64   **INT. LIVING ROOM, ARMSTRONG HOUSE - HOUSTON, TX - SAME TIME**   C64

Mark sits with Janet's MOTHER, reading a book. In the b/g, we
can hear faint comms from the squawkbox.

64   **INT. BATHROOM, ARMSTRONG HOUSE - HOUSTON, TX - DAY**   64

CLOSE ON Janet. Drying her hands, staring into a MIRROR. By
now we know she's not a worrier. But even she is SWEATING.

The comms continue, faint in the b/g, as Janet **STEELS** herself
then opens the door and walks out into --

JIM: They got up to 6 g's during the first stage and 8 g's during the second, so I think intense is the right word. Also, most astronauts noted the "symphony" of mechanical sounds. In other words, the engine noise was pretty loud. Finally, virtually every Gemini astronaut, notably Wally Schirra on Gemini VI-A, commented on the spectacular and slightly disconcerting experience of staging.

JOSH: Thus "Wally's fireball."

JIM: Yes. The fireball—a black-tinged cloud of orange and red flames—was due to the fact that the Titan's second stage ignited while still attached to the first stage, so the exhaust actually rebounded off the top of the first stage.

JOSH: And there was another good jolt for the astronauts when the second-stage engine shut down and they were tossed into microgravity. I first really saw this watching Scott Kelly's launch en route to the International Space Station in March 2015.

JIM: You were a couple drafts in by then.

JOSH: Yep, but still quite a way from the shooting script. I bet you never thought you'd have to give comments on so many drafts.

JIM: It was my pleasure. You and Damien really knew how to treat a book author with respect.

"TO PUT THE AUDIENCE INSIDE THE COCKPIT AND NOT OUTSIDE THE COCKPIT SO THAT YOU CAN FEEL WHAT IT WAS LIKE... THAT WAS ONE OF OUR GOALS."

MARTY BOWEN, PRODUCER

65   __INT. HALLWAY/LIVING ROOM, ARMSTRONG HOUSE - CONTINUOUS__   65

Janet walks down the hall in silhouette to... THE LIVING ROOM.
Rick hovers by a NASA SQUAWKBOX and Mark sits with Janet's
MOTHER (late 50's) in the b/g.

Other than that, it's SURPRISINGLY EMPTY, just a Public
Affairs Officer from NASA and Life Photographer **RALPH MORSE**.

                    JANET
          You need anything, Mom?

Her mother shakes her head. Janet sits on the couch beside
the squawkbox, Morse snapping photos. Janet forces a smile
for the camera...

...but we can see her NERVES as she leans forward, turns up
the volume on the squawkbox.

                    PAO (ON SQUAWKBOX)
          This is Paul Haney in Gemini
          Control, Houston. Our big
          rendezvous display chart here in the
          Control Center shows Gemini 8 in
          orbit. The crew will now attempt to
          find the unmanned Agena spacecraft
          and dock with it.

                                             INTERCUT WITH:

67-68 **OMITTED**   67-68

69-71 **OMITTED**   69-71

72   __EXT./INT. GEMINI VIII, ORBITAL SPACE - DAY/NIGHT__   72

The thrusters fire as Neil begins one of the many burns needed
to rendezvous with the Agena.

Inside the craft, we see the CONCENTRATION on Neil's face as
he gently presses on the **MANEUVER CONTROLLER**...

A beat, then Neil kills the burn.

                    NEIL
Burn end.   DAVE SCOTT
                              Good burn.

Neil's eyes **TICK** over the instruments, glance out the window.

"IT'S NOT ABOUT JUST THE MISSION. IT'S NOT ABOUT JUST GETTING TO THE MOON. IT'S NOT ABOUT JUST THE SPACE PROGRAM. IT'S ABOUT HIM AS A HUMAN BEING AND WHAT IT MEANS FOR A HUMAN BEING TO MAKE SUCH EXTRAORDINARY STRIDES FOR HUMANKIND AND WHAT PUSHES THEM TO PUT THEIR LIFE AT RISK FOR THE REST OF HUMANITY."

CLAIRE FOY, 'JANET ARMSTRONG'

JOSH: One of the things we wanted to highlight here is how isolated the wives were.

JIM: Yes, other depictions don't make this strong enough, not nearly. There really wasn't much in the way of support for the wives during these missions. And the few individuals that were with a wife during the mission—in this case, a *Life* magazine photographer and a NASA PAO—were typically more of an annoyance than a help.

JOSH: So a bit on how these scenes came to be. First off, we had to decide how to condense rendezvous and docking, which involved three burns based on calculations sent up from Mission Control and then three done by Neil and Dave themselves. Frank Hughes was a huge help here. Frank, who served as a Simulation Supervisor or "Sim Supe" at KSC from 1966-71 and rose to Chief of Flight Training in the 90s, walked me through the mission and helped me decide which of the transcript sections to focus on in the script. He also trained our entire team—me, Damien, our cinematographer Linus, our insert unit, and our actors Ryan and Chris Abbott (Dave Scott)—on our Gemini console. He basically taught us how to fly the thing.

That took care of the scenes in the cockpit. But as I mentioned before, we had no comms for Mission Control (outside of what CapCom Jim Lovell says directly to the crew). While we have recordings of the flight director's loop for Apollo, the NASA Historian's office had no such recordings or transcripts for Gemini VIII. To make matters more difficult, Damien wanted to be able to shoot these scenes documentary style, meaning that we couldn't fake the dialogue—we needed real dialogue for all the flight controllers in Mission Control.

To figure out how to approximate that dialogue, I sat with Gerry Griffin, who was a GNC flight controller during Gemini VIII. Gerry laid out who all the important players were in Mission Control and then literally talked me through what each of them would have said at the points in the mission we were focused on. Based on this, I then wrote about five pages of dialogue for each of the nine primary flight controllers, dialogue which would overlap and be performed simultaneously, like a play.

JIM: So the dialogue on page 46 of the script is just a fraction of what you wrote for Scene 73.

JOSH: That's right. And that's true for all the Mission Control scenes in the Gemini VIII sequence. We took a Saturday to do a full rehearsal just for these sections. Fortunately, we cast a number of actual flight controllers, and we had some space enthusiasts like Rick Houston (the co-author of *Go Flight! The Unsung Heroes of Mission Control*) playing extras, which was helpful. We also feature Neil's sons, Rick and Mark Armstrong, as well as Mark's son, Andrew (who we feature playing guitar to celebrate docking).

RIGHT: Cinematographer Linus Sandgren on the Mission Control set.

spacecraft down into a position where its orbital inclination— the angle between the plane of its orbit and that of Earth's equator—matched up with the Agena, the unmanned spacecraft with which Gemini was trying to rendezvous and dock.

This burn came at one hour and thirty-four minutes Mission Elapsed Time (1.34 MET), just as Neil and Dave crossed over the Texas coastline for the first time.

JOSH: The comment about overdoing it a little is actually something Neil said after a later burn.

JIM: Right, that was plane-change burn, which came over the Pacific Ocean just before completing a second orbit, at 2.45 MET.

a little imprecisely, as Neil noted. When Neil's gut feeling was confirmed, Mission Control had him add two feet per second to his speed by making another very short burn.

It took a while before the crew could see its target, the Agena. It was hard to do at night. Neil and Dave had radar information giving them range, range rate, and position. At some point they knew they would see the Agena, but they had to be pretty close to it. According to the mission plan, the astronauts would be in the dark throughout the transfer arc to rendezvous. Then at roughly ten miles out, the Agena would go into daylight, at which point it would light up like a Christmas tree. When that happened, the astronauts could make final adjustments visually.

BELOW: The real Jim Lovell, who served as Capsule Communicator (or CapCom) on Gemini VIII.

                              DAVE SCOTT
                    Shouldn't we have a visual on the
                    Agena by now?

                              *NEIL (INTO COMMS)
                    Houston, I think we overdid it a
                    little.

                                                        SMASH TO --

73   **INT. MISSION CONTROL CENTER, MSC - SAME TIME**   73

Welcome to **Mission Control**, *all the latest technology of 1966.*
PUSH BUTTONS, ROTARY PHONES, STATIC MAPS on the big screens.

                         Mission Control Center
                            Houston, Texas

Lovell, at CAPCOM, talks to the Gemini under the watchful eye
of Flight Director **JOHN HODGE**, 37, British.

          LOVELL (INTO COMMS)
Roger 8, stand by for a   HODGE (INTO HEADSET)
correction.   Fido, Flight, how are we
                                       doing?

Hodge looks to FIDO.

                              FIDO (INTO HEADSET)
                    We've got solid track on both
                    vehicles, calculating now.

Kraft watches nearby, along with Ed, Gus, the other astronauts
and a few USAF OBSERVERS, including a BLACK PILOT, 31. We
note his name tape, **R. LAWRENCE.**

                              AGENA CONTROL (INTO HEADSET)
                    Fido, Agena, you have what you need
                    from us? Fido, Agena?

          HODGE (INTO HEADSET)
Fido, did you copy that?   GUIDANCE (INTO HEADSET)
                                       You guys getting this?

GUIDANCE shows a printout to Fido.

                              FIDO (INTO HEADSET)
                    Roger, flight, I copy. We just have
                    some ratty data from the Gemini
                    computer.

Fido, Guidance and Retro huddle. Hodge grows impatient.

> HODGE (INTO HEADSET)
> I need the correction,   GUIDANCE (INTO HEADSET)
> gentlemen.   Sending it up now.

Fido writes up the PAD, hands a copies to Lovell and a tech
who runs into the projection room. It APPEARS on screen.

> HODGE (INTO HEADSET)
> Okay, capcom, let's get it up to
> them.

> LOVELL (INTO COMMS)
> Gemini 8, Houston capcom. We want
> to give you another burn here very
> shortly. Stand by to copy. GET B:
> 03:03:41; Delta-V is 2 seconds; 2
> feet Posigrade...

74   **OMITTED**   74

75   **INT. ARMSTRONG HOUSE - DAY**   75

CLOSE ON the squawkbox. Janet sits on the couch leaning over
it, listening intently.

> LOVELL (ON SQUAWKBOX)
> *...8, Houston. Do you copy?*

No answer; it's <u>unnerving</u>. More so, when Mark runs in and
grabs the squawkbox.

> JANET
> Mark, give that back. Mark, give
> that back, put it back on the table.
> I'm not joking, Mark.

The LIFE photographer begins snapping photos.

> LOVELL (ON SQUAWKBOX)   JANET
> *Gemini 8, Houston capcom.*   Honey, give me it, it's
>   really important. Give that
>   back to mommy right now.

> LOVELL (ON SQUAWKBOX)   JANET
> *8, do you read? Copy.*   Mark Armstrong, if you don't
>   give me that back...

Mark refuses to put it down. He smiles. Janet, trying to
control herself, bends over Mark. She LOWERS her voice.

> LOVELL (ON SQUAWKBOX)
> *8, can you give us a status?*

JOSH: The dialogue in Scene 75 was actually all improvised on the day by Claire Foy (Janet) and Connor Blodgett (Mark Armstrong). We loved how human it was, the surreal juxtaposition of the quotidian and the extraordinary. And while this wasn't based on a particular anecdote, both Mark and Rick emphasized how "hyper" Mark was as a kid. This seemed to fit that description.

MIDDLE LEFT: Andrew Armstrong on set in Mission Control.
MIDDLE RIGHT: Neil's sons, Rick and Mark Armstrong, in action on set in Mission Control.
BOTTOM: Janet tries to hide her concern during Gemini VIII, while *Life* photographer Ralph Morse (Dustin Lewis) tries to capture it.

                         JANET
                 I'm not joking, Mark.

Off Janet, frustrated, **SMASH TO --**

76   <u>**INT. GEMINI VIII COCKPIT - NIGHT/DAY**</u>   76

Neil checks the rendezvous chart as Dave stares at the
computer read out; <u>something's off</u>.

          DAVE SCOTT
I'm getting a horrendous 20   NEIL
to 25 feet per second down,   I can't see any possible
Neil.   reason for that.

          DAVE SCOTT
Where are we on the plot?   NEIL
                                We're up above it.

          DAVE SCOTT
Right, but what does it look   * NEIL
like if --   I can't-- I'm sorry, I have
                                to, I have to look at this...

Dave QUIETS as Neil keeps working the numbers.

     LOVELL (COMMS)
8, can you give us a status?   NEIL (INTO COMMS)
                                No, I've got too much to do.

     LOVELL (COMMS)   NEIL
Copy. Standing by.   (beat, to Dave)
                                ...okay. We're going to go
                                with the closed loop. 25
                                forward, 8 left, 3 up, and
                                I'm going to RATE COMMAND.

Dave quickly adjusts the dials. A beat, then Neil's eyes **TICK**
to the clock.

                         NEIL
                 ...3, 2, 1, burn.

JOSH: The language at the top of Scene 76, like most of the comms, is verbatim from the TEC transcript. I loved this language because it underscores how challenging rendezvous was and it illustrates how Neil could/would be polite but direct when he was focused on a problem.

JIM: All rendezvous maneuvers in space are pretty fascinating. In this case, the Gemini was in a circular orbit 15 miles closer to Earth than the Agena, so the spacecraft was moving faster and catching up to the Agena. When the angle was just right, Neil and Dave initiated the transfer arc, slowly ascending while keeping the target in sight. Along the way, they made two mid-course corrections in order to hold the target fixed in position against the field of stars. Then, when they were within a mile or so, they started braking and came to a halt alongside the Agena. What's clever is that, although they were flying a curve, by holding the target fixed relative to the star field, they actually made a straight-line approach.

JOSH: Obviously, we're condensing time and omitting a fair amount of the action and chatter. But hopefully, the audience will still get a sense of the magic of all this.

We can't overstate how crucial the help from Frank Hughes was. For example, Neil calls for a digital range and rate as he brakes. At first, we didn't understand this because Neil had (analog) range and rate gauges on his side of the Gemini console. Frank explained that Neil would ask Dave for range and rate because (1) the rudimentary computer on Dave's side of the console gave more precise information and (2) Neil had to keep his eyes on the Agena and the transfer arc plot (which specified what his range and rate should be).

JIM: I love how excited Neil and Dave get when they come up on the Agena.

JOSH: Again, this is verbatim from the transcript.

Neil **HITS** the thrusters. The craft **SWINGS** left and Neil lets
go. Dave eyes the console, jots down residuals...

...then spots Neil staring. Dave follows his gaze to what
looks like a BRIGHT STAR in the window.

            NEIL
Could be a planet.
                                    DAVE SCOTT
                              Could be.

It's not. Dave **SMILES**.

            LOVELL (COMMS)
This is Houston, we have your
ground TPI backup when you're   DAVE SCOTT (INTO COMMS)
ready to copy...   Stand by. We have a visual
                                on the Agena... at least we
                                have something we think looks
                                like the Agena.

               LOVELL (COMMS)
          Understand, possible visual on the
          Agena.

Neil SQUINTS at the Agena **GROWING** in the window as the sun
rises over the earth.

            NEIL
We're getting a little Out-Of-
Plane now...   DAVE SCOTT
                                   (checking the computer)
                                We've got to get 3 aft and 2
                                1/2 up...

Neil sets braking thrusters, grips the maneuver controller.

             *NEIL
I'm going to start braking. Give me
a digital range and rate.

Neil hits the **BRAKING THRUSTERS**. The Gemini slows as it
rushes to meet the Agena.

            NEIL   DAVE SCOTT
I'd better back off a bit.   (takes a reading)
                                6,000 feet, 31 feet per
                                second.

Neil hits the braking thrusters again.

            NEIL
Put in a little to the left.

            DAVE SCOTT
1680 feet.

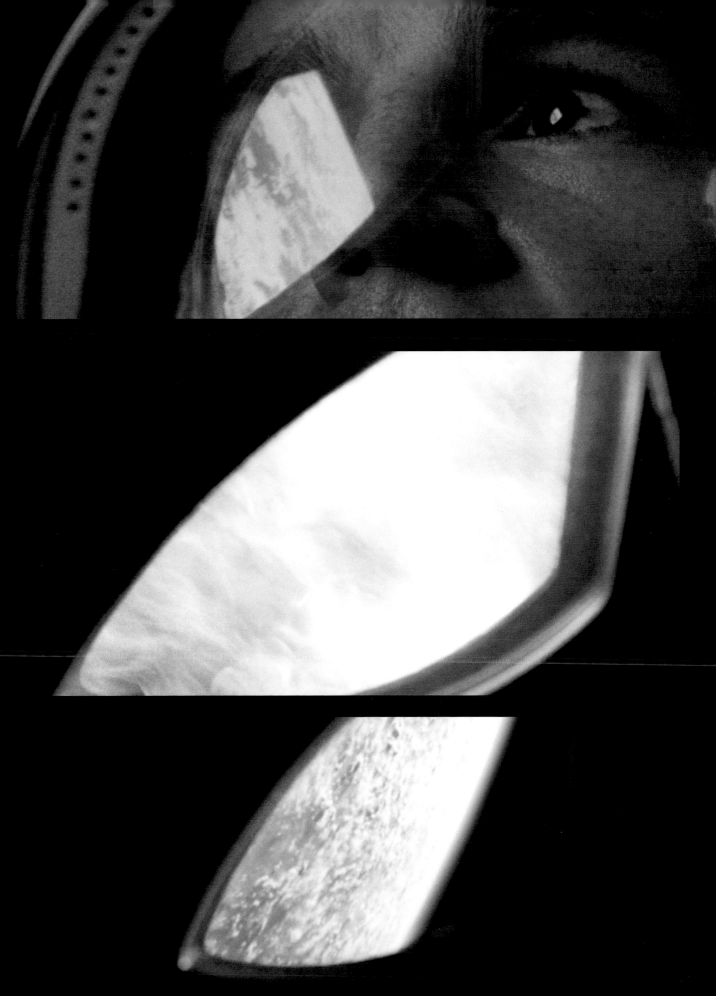

And now the Agena **GROWS** in the window, until it's HOVERING in full view. Neil smiles.

> NEIL
> That's unbelievable.

> DAVE SCOTT
> Would you look at that!

Dave smiles broadly, Neil smiles back. An unusual TWINKLE in Neil's eyes and, again, a touch of that **CHILDLIKE WONDER.**

> LOVELL (COMMS)
> *Gemini 8, Houston. Standing*   DAVE SCOTT
> *by for rendezvous remarks.*   You tell them.

> NEIL (COMMS)
> *Houston, we're station-keeping on*
> *the Agena at about 150 feet.*

77   **INT. MISSION CONTROL CENTER, MSC - DAY**   77

CLOSE ON Ed, tense... breaking into a smile.

He looks over at Gus and a few of the astronauts, excitement and relief. Lots of smiles. **RACK TO** Hodge at a console.

> HODGE
> Okay, stay focused gentlemen, we're
> only halfway there.

Fido turns to Retro.

> FIDO
> Thanks for the extra hands.   HODGE
>   Stay focused, gentlemen.

The men turn back to their consoles as we **CUT TO --**

EXT. GEMINI VIII, ORBITAL SPACE - SAME TIME

From a distance, we see the Gemini and the Agena, two tiny objects just below the earth.

78   **INT. GEMINI VIII - DAY**   78

Dave is smiling, still feeling the rush. He looks to Neil. Who's smiling as well...

> NEIL
> Man it flies easy.

First Man   POST CONFORMED BLUE   50.

A79   <u>**EXT. GEMINI VIII, ORBITAL SPACE - SAME TIME**</u>   A79

CLOSER ON the Gemini and the Agena, now moving towards each
other in a graceful ballet. It seems effortless, a joy to it.

As the Gemini inches closer to the Agena, we **INTERCUT WITH --**

B79   <u>**INT. GEMINI VIII - SAME TIME**</u>   B79

Neil flies the Gemini, a **LIGHTNESS** to him. The grit of
training, the darkness of Elliot's death have all **FALLEN AWAY.**

> DAVE SCOTT
> Does it really?

> NEIL
> This station keeping, it's just,
> it's like nothing.

Dave SMILES at Neil's enthusiasm. Outside the ship, we see
the Gemini has crept up right beside the Agena.

> DAVE SCOTT (INTO COMMS)
> RKV, this is 8. We're
> sitting about 2 feet out.

> RKV CAPCOM (COMMS)
> *Roger. Stand by for a couple
> minutes here.*

> DAVE SCOTT (COMMS)
> *Roger.*

HOLD ON Neil and Dave, waiting on the precipice.

> RKV CAPCOM (COMMS)
> *Okay Gemini 8, we have T/M solid.
> You're looking good on the ground,
> go ahead and dock.*

Dave enters '221' on the Agena encoder. We hear a **BUZZ.**

Neil lines up the GEMINI DOCKING BAR with the SLOT on the
Agena... then squeezes the throttle, moving forward SLOWLY.

Outside the ship, the Gemini and the Agena **CRUNCH** together.

> NEIL (INTO COMMS)
> Okay, I'm going to cycle our
> Rigid/Stop switch now.

Neil nods to Dave. Moment of truth. Dave hits a switch. We
hear the motor aboard the Agena WHIR and we **PUSH OUTSIDE.**

First Man   POST CONFORMED BLUE   51.

79   **EXT. GEMINI VIII, LOW EARTH ORBIT - DAY**   79

From afar, we see the Agena **CLASP** onto the Gemini... There's a
loud **CLANK**. **SMASH BACK INTO --**

80   **INT. GEMINI VIII COCKPIT - DAY**   80

In the cockpit, the 'RIGID' button on the Agena Station
display panel **LIGHTS UP**. Neil and Dave share a **LOOK**.

Neil puts out a hand. Dave smiles broadly, shakes it.

> NEIL
> Flight, we are docked.

81   **INT. MISSION CONTROL CENTER, MSC - DAY** (5:10 PM)   81

Ed **SMILES BROADLY** as the other astronauts in the room explode
with **CHEERS**. Kraft shakes Hodge's hand. Gus calls out...

> *GUS
> Someone call Cronkite, have him tell
> the Soviets they can go screw! And
> Pete, call those idiots in Congress
> while you're at it!

More CHEERS, laughter as Deke, Conrad and Dick Gordon ENTER in
FLIGHT SUITS (from the Cape). Deke smiles pleased, as the men
slowly get back to work.

> FIDO (COMMS)
> *Okay, let's go ahead and get a state
> vector for the combined spacecraft.*

82   **INT. ARMSTRONG HOUSE - DAY**   82

Morse takes a photo of Janet and the kids.

> MORSE
> Congratulations, Mrs. Armstrong. A
> great day for the United States.

The kids jump up. Janet forces another smile. Still uneasy.

> RICK
> Hey mom, can I go to Randy's house?

> JANET
> Sure you can. You just have to be
> back by 7, okay?

Rick runs off.

JOSH: Docking was a huge achievement—and the first time we beat the Soviets on such a major achievement in the Space Race. So it certainly felt like some celebration was called for. When you read the transcript, you can hear the excitement from the crew. While we don't use the line, Dave confirmed via Mark Armrstrong that he'd happily told Houston the dock was "really a smoothie."

JIM: I love the reference to Walter Cronkite, who not only served as anchorman for the CBS Evening News at the time, but also spearheaded CBS's extensive coverage of the U.S. space program. Cronkite was so respected by NASA that he became the only non-NASA recipient of an Ambassador of Exploration award.

ABOVE: The view of the Agena from Gemini VIII prior to docking.

> **"WE USE LED LIGHTS IN A WAY SO THAT WE DIDN'T HAVE TO USE GREEN SCREEN. WE COULD MAKE THE ACTORS REALLY FEEL LIKE THEY'RE THERE IN SPACE."**
>
> WYCK GODFREY, PRODUCER

JOSH: Gemini VIII is one of my favorite set pieces in the movie, in large part because very few people recall the mission, let alone that Neil and Dave almost died. What's particularly scary is that the problem surfaced while Neil and Dave were out of contact with Mission Control. When they regained contact, Mission Control couldn't offer much help. For all intents and purposes, Neil and Dave were on their own.

JIM: Lovell's dialogue, which is condensed from the transcript, sets this up well (especially as you have him say "loss of signal" instead of using the acronym "LOS" as he did at the time). During Gemini VIII, Houston had to rely upon a worldwide tracking network with stations at various spots around the globe to relay comms up to the orbiting spacecraft. While this worked relatively well, there were "dead zones" between tracking stations when the spacecraft would lose the signal entirely.

JOSH: This particular dead zone lasted twenty-one minutes; it was during this loss of signal that the Gemini began to spin.

JIM: The TEC transcript does not include any of the on-board comms for the period when Neil and Dave are in this dead zone.

JOSH: As far as we could tell, there is no transcription or recording of these comms. So all dialogue after LOS had to be reconstructed. We did this based on descriptions of what happened during the spin as well as Neil and Dave's March 26, 1966 Pilot's Report Press Conference. We also went through this section with Frank Hughes, who walked us through everything Neil and Dave would have attempted in order to right the ship.

JIM: When Gemini VIII started to spin, the Agena docking satellite was the prime suspect. As Lovell suggested just before LOS, there had been many problems with the Agena during development, so it was logical to conclude that any major problem would come from the Agena.

JOSH: Neil and Dave spent a good deal of time trying to correct the spin while still docked to the Agena. We compress time for the sake of the drama, but they cycled the Agena multiple times, with some success at first. When these attempts failed, Neil concluded that the Agena was the problem and decided to separate from the Agena.

JIM: Only once they had undocked from the Agena and the problem worsened did they realize the issue was with the Gemini. After the fact, they determined that a short circuit had stuck open thruster eight on the ship's Orbit Attitude and Maneuvering System (OAMS). But there was no way for Neil or Dave to know that at the time as they would have only heard the thruster when it fired, not when it was running steadily.

> > RICK
> Yes, ma'am.
> >   MARK
> >   Can I go too?

Mark calls out as Janet grabs a coffee cup.

> > JANET
> No, you're staying here with me.

> > MARK
> Nooo.

> > *LOVELL (ON SQUAWKBOX)
> *Gemini 8, we're about to have loss
> of signal, we'll pick you up over
> the hill for Dave's EVA.*

83   **INT. GEMINI VIII COCKPIT - NIGHT**   83

Dave grabs a pen, takes notes as Lovell continues.

> > LOVELL (COMMS)
> *...ENABLE the SPC's... if you run
> into... Attitude... Agena...*

But the comms turn **STATIC** as they lose signal and go into the
dark, earth night. Dave drops his pen, grabs a book but sees
Neil's pulled out two PACKETS. ***DAY 1, MEAL B.***

He holds a packet to a nozzle on the console, injects it with
water. It looks pretty bad. Analytical --

> > NEIL
> Man, that's peculiar.

Dave, who's been looking through a manual, turns to him.

> > DAVE SCOTT
> Oh, great.

> NEIL
> I think there's some air   DAVE SCOTT
> bubbles in it.   I think I'm gonna save mine
>   for later. A little treat.

They laugh. Dave puts his meal aside and reaches for a
manual, glancing at the console... PAUSING.

> > DAVE SCOTT
> ...Neil, we're in a bank.

Neil's eyes **TICK** to the **8 BALL**. It shows a 30 degree roll.

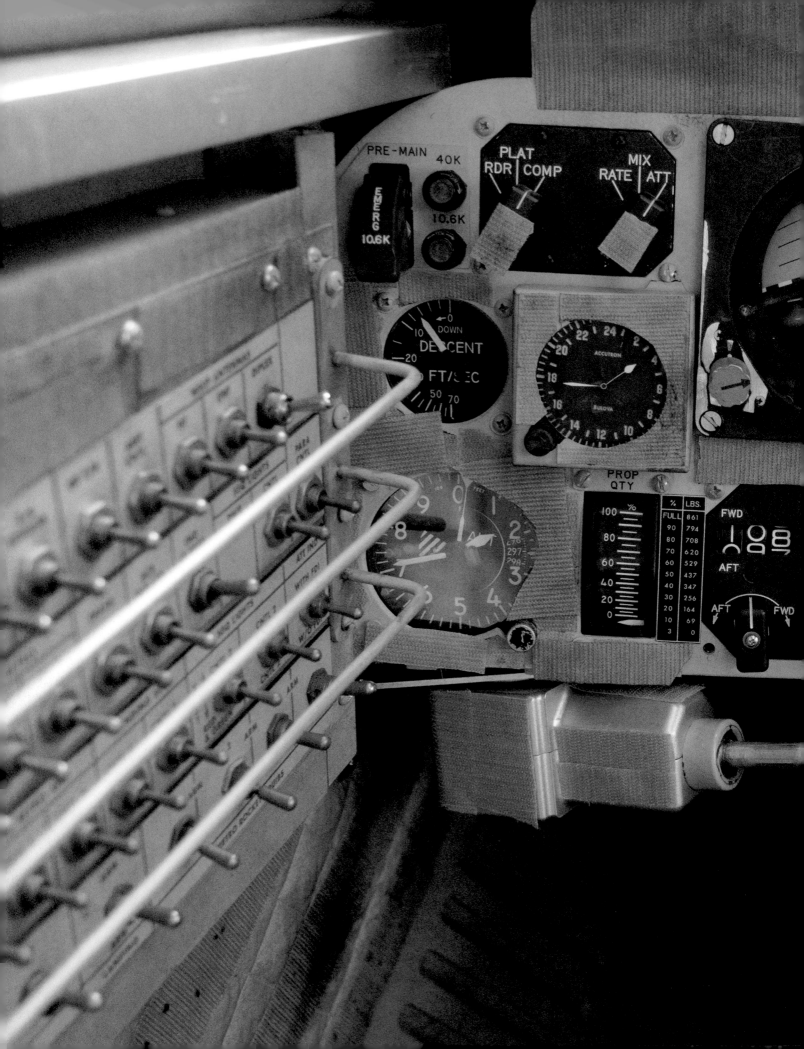

"YOU START WITH DIAGRAMS AND
DASHBOARD INSTRUMENTATION
AND YOU END UP TALKING WITH
EXPERTS, AND ENTHUSIASTS, AND
NASA, AND YOU TRY TO ANALYZE IT
ALL. IT'S LIKE BEING A DETECTIVE."

NATHAN CROWLEY, PRODUCTION DESIGNER

> DAVE SCOTT
> We're not doing it, it's not
> us, it must be the --

> NEIL
> Shut off the Agena's control
> system.

Dave quickly follows orders, punching in commands to turn off
the Agena ACS, horizon sensors and geocentric rate.

> DAVE SCOTT
> Code 400, Agena control system is
> shut down.

Neil hits the stick, WATCHES the 8 ball... it isn't working.
He switches to rate command, keeps trying...

...when the items floating in the cabin are **SLAMMED UP** against
the walls, as if by an unseen force.

Neil moves to direct. Still can't control the spin. Shit.
Neil scans the console, turns to Dave...

> NEIL
> Cycle the Agena.

> DAVE SCOTT
> (commands encoder)
> Turning it on...
> (commands encoder)
> ...turning it off.

...but the items remain pressed to the walls. The sun rises
and now the earth swings past the window OVER AND OVER. We
realize **WE'RE SPINNING**. It's **DIZZYING**.

> DAVE SCOTT
> I'm gonna cycle the ACME and the
> propellant motor valves.

Dave does this, no improvement.

> NEIL
> Switching ADL to pitch.
> (no response)
> RL to pitch.

PUSH IN on Neil, working the problem, eyes **TICKING**...

...to the **8 BALL** showing a HUGE BANK...

...to the **ROLL RATE GAUGE** moving past **180°/SEC**...

...to **DAVE** STRUGGLING with the disparate G-forces...

> DAVE SCOTT
> Roll rate is at 180... 190...

First Man   POST CONFORMED BLUE   54.

                        NEIL
            Separate from the Agena.

Dave hesitates, but Neil's not waiting. Dave gets to it.

                        DAVE SCOTT
            Setting Agena to allow remote
            command, switching on the DAC.

Dave quickly punches commands into the encoder.

                        DAVE SCOTT
            Make sure you give it extra
            thrust so we don't smash into   NEIL
            the Agena --   On my mark, undock.

The RIGID LIGHT goes **OUT**. Neil **GRABS** the maneuver thruster.

                        NEIL
                2, 1...

Dave **FLIPS** the undock switch; Neil **PULLS** back on the maneuver
thrusters to pull away and the Gemini **JERKS** back...

The Agena SPINS VIOLENTLY outside the window, **NARROWLY MISSES**
the nose of the Gemini. Dave looks relieved... until he feels
the **G FORCES**. He looks out the window, sees the world pass.

Shit. *They're spinning even faster now*.

Neil keeps hitting the throttle... but it doesn't help.
SURPRISED, he drops the stick, eyes **TICKING** from the console
to Dave...

                        DAVE SCOTT
            OAMS propellent down to 13 percent,
            our roll rate is still rising...
            It's not the Agena, it's us!

Neil **STRAINS** against the **ROCKETING G-FORCES**, NAUSEATING and
DEADLY... his eyes **TICKING** to the roll rate gauge at **300°/SEC**.

Off the two men, **STRUGGLING** to work the problem, **SMASH TO --**

84 **INT. HOLD, COASTAL SENTRY QUEBEC, PACIFIC OCEAN - SAME TIME** 84

A windowless room in the hold of the **U.S.N.S. SHIP**. **JIM
FUCCI**, 42, mans a state of the art TRACKING CONSOLE.

        MCC NETWORK (COMMS)
Gemini 8 coming back into   FUCCI (COMMS)
range in 3, 2, 1...   This is CSQ checking our
                                    commlink. How do read?

JOSH: We take some license here. The speed at which everything moves to the cabin walls is heightened. The speed at which Earth moves through the windows is heightened too. The lights shorting out is a fiction. But this is all to put the viewer in the cockpit, to convey how intense and dangerous this situation was.

JIM: It's a corollary of Murphy's law that bad things always happen at the worst possible times. As we've discussed, Neil and Dave were out of radio contact almost the entire time of their emergency and, once they regained contact, it was over Coastal Sentry Quebec, a tracking ship at sea with limited ability to communicate or transmit data to Houston.

JOSH: This resulted in a precarious game of telephone. We found the transcript in the *New York Times* archives. The conversation between the Gemini, Jim Fucci (the CSQ CapCom), and Mission Control was published on March 18 and it's fascinating.

JIM: It highlights not only how limited the technology was, but also how hard it was for Mission Control to understand what the problem was, let alone help.

JOSH: As I mentioned earlier, in this most precarious of moments, Neil and Dave were truly on their own.

One other interesting note about Scene 84. In the film, the cut pattern at the end of the scene is a bit different from what's scripted. We filmed all of this dialogue in the spacecraft set, on our CSQ set and with Claire and the squawk box at the Armstrong House. Damien did this to give us more options in post, so we could play with how to best get across the increasing level of tension. I love the unscripted cutaway to Janet in the movie—truly harrowing.

Nothing.

   FUCCI (INTO COMMS)
Gemini 8. How do you read?   NEIL (COMMS)
             We have serious problems.

Fucci **FREEZES**, stares at his instrument panel. The men around
him go quiet...

   NEIL (COMMS)
We're, we're tumbling end
over end up here, we're
disengaged from the Agena.   FUCCI (INTO COMMS)
           Okay. We got your spacecraft
           free indication here... what
           seems to be the problem?

     NEIL (COMMS)
..we're rolling up and we can't turn
anything off. We're continuously
increasing in a left roll...

   HODGE (COMMS)
CSQ, Flight.   FUCCI (INTO COMMS)
           Go ahead, flight.

Off Fucci, PALE, **SMASH TO --**

85 **INT. MISSION CONTROL CENTER, MSC - EVENING**   85

Kraft, Deke, Hodge and Ed stand by Lovell. The normal chatter
in the room has all dropped off.

    *HODGE (INTO COMMS)
   Did he say he could not turn the
   Agena off?

Everyone is quiet, listening intently to --

     FUCCI (COMMS)
No, he says he is separated from the
Agena and he's in a roll and he
can't stop it. It's approaching one
revolution per second, at that rate
they could black out any minute...

Ed **TIGHTENS**, looks to the others.

    DEKE
Paul. Paul.

Deke gets the attention of PAO Paul Haney, motions for him cut
the public feed. As he does, we **SMASH TO --**

JIM: This is a true story, one Janet left out of her initial interviews in the late 60s. NASA indeed shut off her squawk box and Janet insisted that the NASA PAO assigned to the Armstrong home drive her to Mission Control.

JIM: The Gemini's rate of rotation kept increasing until it reached the point where the motions began to couple. In other words, the problem became not just a precariously high rate of roll but also the coupling of pitch and yaw. The spacecraft basically turned into a tumbling gyro.

JOSH: A real life version of the MAT trainer we saw earlier. But rotating much faster. In fact, Dave Scott told us the centrifugal force pushed everything in the cabin up against the walls.

JIM: It's pretty spooky when it happens in the film. Worse, as the revolutions approached 360° per second, the astronauts began to lose peripheral vision.

JOSH: They were in serious danger of losing consciousness. It was the first potentially fatal in-space emergency for the American space program.

JIM: The fact that it wasn't fatal speaks to Neil's ability to remain cool under pressure and work the problem. Unable to stabilize the ship even after shutting everything down, Neil realized the only way to stop the spin was to engage its only other control system—the Re-entry Control System (RCS), which was designed to control the spacecraft as it reentered Earth's atmosphere.

JOSH: Even this was challenging, as it required closing the two RCS breakers overhead before turning the RCS switches on the front dash to RCS direct.

JIM: Only someone with intensive training would be able to find those overhead breakers in the tumbling craft. Of course, the first time Neil tried this it didn't work as they'd switched the ACME power bias off while docked. Just another layer of problem solving.

JOSH: One interesting note here—Damien and I had always liked the suspense you get in cutting away from Neil and Dave at the end of Scene 88 before they right the ship. But when Steven (Spielberg) saw the cut, he suggested that it might actually be more intense if we stayed with the astronauts until they get the spin under control. Damien tried it and loved it. So this how we have it in the film. I left it as we had it in the script because I like how it reads; but on film, there's no question it has more impact the other way.

There's an old line about how every movie is written three times—at script, in production, and in post. I certainly believe that to be true. To me, it's like a coloring book that you fill in as you shoot the film. Individual scenes and sections of the film feel different as you shoot them and certainly feel different once you start piecing together all the footage. Moments that you were convinced would work at script stage don't work in the cut; and vice versa. This is why a script really continues to evolve until you lock picture and sound.

BELOW: Linus Sandgren in the cockpit.

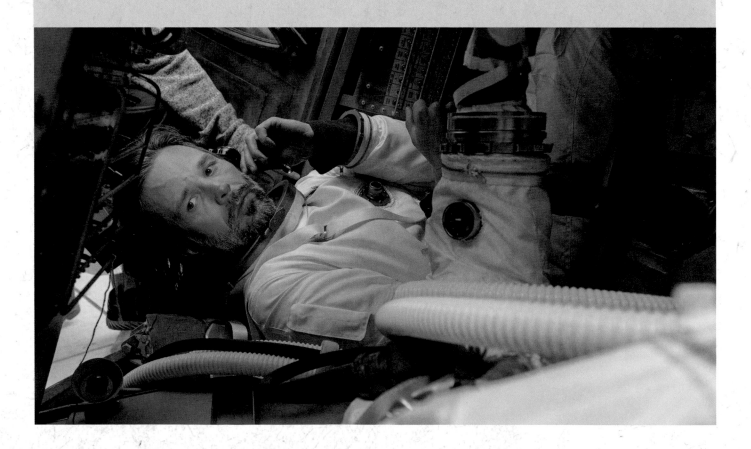

First Man   POST CONFORMED BLUE   58.

> HODGE (COMMS)
> *CSQ, is there a status update?*

> *FUCCI (COMMS)
> He's blown both RCS squibs. They
> have initiated squibs and blown 'em.*

Why blow RCS squibs? An awful beat as the men try to process.

> FUCCI (COMMS)
> *And he's lost considerable gas
> pressure in...*

STATIC cuts him off. Then, faintly...

> NEIL (COMMS)
> *Okay we're, uh, regaining control of
> the spacecraft slowly in RCS direct.*

A huge intake of breath.

> FUCCI (COMMS)   HODGE (INTO HEADSET)
> *Roger, copy.*   (relieved)
>   Roger, copy.

> NEIL (COMMS)
> *We're pulsing the RCS slowly, it's
> all roll right...*

We clock the **RELIEF** in Ed's eyes as we **SMASH BACK TO** --

91   **INT. GEMINI VIII COCKPIT - DAY (SAME TIME)**   91

Dave, foggy, unsettled, watches Neil calmly pulsing the RCS,
his eyes on the DROPPING roll rate gauge: *90°... 80°... 70°.*

> NEIL (INTO COMMS)
> We're trying to kill our roll rate
> here.

Dave nods. The sun stops strobing, flight plans and ephemera
peel off the walls and start floating about the cabin again...

> NEIL   DAVE SCOTT
> Move us back to one ring.   Copy.

Off Neil, **CUT BACK TO** --

92   **INT. MISSION CONTROL CENTER, MSC - NIGHT**   92

Kraft leans into Hodge, Deke hovering.

STAY INT. GEMINI

JOSH: The Gemini needs RCS fuel to orient the ship on re-entry. Without proper orientation the ship would burn up. Hence the immediate concern about RCS fuel.

JIM: It's also why, after the fact, a few astronauts criticized Neil for pressurizing and using both RCS rings. The rings are redundant and these astronauts thought that if he'd only pressurized one ring the mission could have continued.

JOSH: Which was incorrect. NASA's mission protocol clearly stated that once either RCS ring was blown the mission had to be aborted. We have Kraft consult Gilruth to keep the character alive, but there wasn't much gray area here. So any notion that the mission could have continued after Neil hit that switch (or switches) is incorrect.

JIM: That's right. Further, my understanding is that there was no way for Neil to pressurize just one of the rings, because the system didn't allow for that. Activating either RCS switch would pressurize both rings.

Of course, this took us a while to figure out, which just goes to show that understanding all of the technology, instrumentation, and procedures involved in a Gemini or Apollo mission is not easy even for professional engineers, which we aren't.

JOSH: The instrument panel we used for Gemini VIII was built by John Fongheiser, a hobbyist who's been fascinated by the Gemini missions for decades and created his own company (Historic Space Systems) to build historically accurate spacecraft exhibits for museums. Our production designer Nathan Crowley managed to find John's company online and he offered to let us use his Gemini instrument panel—which was both historically accurate and operational. After years studying diagrams, it was a real thrill to be able to interact with such a terrific facsimile.

First Man   POST CONFORMED BLUE   59.

                         *KRAFT
              I want emergency landing options.

                    DEKE
              You don't wanna wait to find out how
              much fuel he's got left?

Kraft considers, then turns to Gilruth.

                    KRAFT
              Bob, what do you think?

                    GILRUTH
              I think they'd better land now.

Deke **PROCESSES** as Ed walks up. He leans in, sotto --

                    ED
              Deke. Jan's outside.

93 **INT. HALLWAY OUTSIDE MISSION CONTROL CENTER, MSC - NIGHT**   93

CLOSE ON Janet by the door. **CLENCHED.** Eyes radiating INTENSE
RAGE. And **TERROR.** A beat, then Deke and Ed walk out.

                    DEKE
              Jan, the ship is stable, they're
              going to be all right, Jan.

Janet doesn't believe him. She looks to Ed again.

                    ED
              He's okay, Jan.

She takes a breath, processing.

                         *DEKE
              I need you to go home.

                    JANET
              Fine. Turn the box back on.

          DEKE   JANET
I'll see what I can --   Now. Turn the box back on
                                   now.

          DEKE   JANET
...well, there's security   I don't give a damn. I've
protocols --   got a dozen reporters on my
                                   front lawn, you want me
                                   telling them what's going on?

JIM: As I said earlier, Janet had the NASA PAO drive her to the Manned Spacecraft Center. There, she was stopped at the front door of the control center. As she said, "I was denied entrance and I was furious."

JOSH: We're not sure who came out and talked to Janet. Deke and Ed likely would have been too busy with the emergency. But, Janet did talk to Deke later about the situation and told him (in no uncertain terms) not to keep her in the dark again.

JIM: She said something to the effect of, "If there is a problem, I want to be in Mission Control and if you don't let me in, I will blast it to the world!"

JOSH: For the sake of time, we conflated those two incidents.

And Claire suggested including a version of the Janet quote. All of which felt in keeping with the spirit of what happened. Plus, seeing Janet get that hot with Deke (while Ed stands listening just outside Mission Control) felt like the best way to get across how furious she really was and how harrowing this must have been. Frank Hughes got to know Janet around this time. After seeing the movie, he said that Claire was terrific—but that she maybe wasn't angry enough in this scene!

Moving on to Scene 97, almost all of the dialogue is right from the TEC transcript, although time has been condensed and a lot of the re-entry prep and accompanying verbiage has been left out. I love the two highlighted lines as they show frustration and a bit of unease from Neil, who was generally unflappable. To me, this demonstrates how dismayed he was at having to head home early.

"COLOR IS USED WHEN WE'RE TRYING TO TELL PARTS OF THE STORY... IT WAS DEFINITELY DRIVEN BY THE EMOTIONAL CONTENT IN THE SCRIPT."

MARY ZOPHRES, COSTUME DESIGNER

First Man   POST CONFORMED BLUE   60.

                         DEKE
          Jan, you have to trust us, we've got
          this under control.

                         JANET
          No, you don't. All these protocols
          and procedures to make it seem like
          you have it 'under control'. But
          you're a bunch of boys making models
          out of balsa wood, you don't have
          anything under control.

TEARS WELL; embarrassed and FURIOUS, she **STALKS OFF**. Off Ed --

94   **OMITTED**   94

95-96 **OMITTED**   95-96

97   **INT. GEMINI VIII COCKPIT - DAY**   97

Dave stows gear while Neil checks headings, helmets now on.

                         CAPCOM (COMMS)
          Naha RESCUE 1 will be on station at
          splashdown with a flotation collar.

The COMMS **CUT OUT**. They're on their own.

                         NEIL
          Did you get the call signs?

                         DAVE SCOTT
          Yeah. It's Naha RESCUE 1, Naha
          SEARCH 1.

Neil pauses.

                        *NEIL
          Well, I'd like to argue with them.
          About the going home. But I'm not
          sure how we can.

                         DAVE SCOTT
          Yeah.

The slightest hint of frustration. But there's nothing to do.

                        *NEIL
          I keep thinking is there anything
          else that we forgot...

                         DAVE SCOTT)
          We did everything, far as I know.

First Man   POST CONFORMED BLUE   61.

                              NEIL
                    Okay.

Dave turns to the basic computer, starts punching in numbers.
Neil prepares as well, when in the window he sees...

The **AGENA.** Floating in the distance. Neil stops, confronting
the FAILURE of the mission. He's quietly devastated.

He sits there for a moment, staring out, then gets back to it.

Neil looks UNSURE. But they need to finish. Neil flicks the
four squibs, they close their visors and...

                         CAPCOM (COMMS)
                    5, 4, 3, 2, 1, retrofire.

                                                   SMASH TO --

98   **INT. AUDITORIUM, MANNED SPACE CENTER - HOUSTON, TX - DAY**   98

Gilruth stands at the podium, addressing a **MASS** of reporters.
Deke watches from the audience.

                 Gemini VIII Pilot Press Conference
                           March 26, 1966

                              GILRUTH
                    Gemini 8 saw two complex vehicles
                    launched on the same day, on time.
                    We saw a flawless rendezvous and
                    docking. All of which has tended to
                    be overshadowed by the malfunction.

Find Neil behind him with Dave and NASA personnel. Neil looks
EVEN MORE UNCOMFORTABLE than he was moments ago...

                              GILRUTH
                    But I think we should focus on the
                    progress resulting from the mission.

99   **INT. CONFERENCE ROOM, MANNED SPACE CENTER - DAY**   99

CLOSE ON a finger pressing **"Record"** on a REEL-TO-REEL.

                         GEORGE MUELLER
                    The board would like to focus on the
                    malfunction.

Gilruth, Kraft, Deke and Nasa Associate Administrator **GEORGE
MUELLER**, 49, at a long table. Neil and Dave sit across from
them. It's a **FORMAL MISSION REVIEW...**

...and judging by Mueller's tone, Neil's job is on the line.

JOSH: The press conference dialogue is all taken from the transcript of the March 26 Pilot's Report Press Conference. The mission review dialogue is based on Neil's comments on the emergency.

JIM: The script seems too dramatic to me here in saying that Neil's job was on the line. NASA wanted to get to the bottom of what happened, but there never was a serious concern that Neil would lose his job.

JOSH: Maybe not, but Neil was concerned. As he said, "I think if it had turned out that we, in fact, had made a mistake—a little one or a big one—that would have been a serious issue. Dave and I couldn't identify serious mistakes that we made, but we recognized that maybe we did make some. So I'm sure there was a concern that it might affect us some way in the future."

JIM: That's true. But despite some Monday morning quarterbacking by a few astronauts, the strong consensus was that Neil and Dave had done absolutely everything that they could have given the situation.

JOSH: And I agree with that. But the story is from Neil's point of view. And from what you've told me, no one was a tougher critic of his performance than Neil himself.

JIM: He always felt if he'd been a little smarter, he might have been able to diagnose the problem and come up with a solution more quickly and avoided aborting the rest of the mission. Of course, I disagree.

JOSH: As do I.

One small note , we included John Hodge (Ben Owen) at the press conference so the audience can see one of the faces from Mission Control. In reality, the only NASA personnel beyond Neil and Dave at the press conference were Bob Gilruth, Dr. Robert Seamans, the NASA Deputy Administrator, and Julian Scheer, the Assistant Administrator for Public Affairs.

ABOVE: The Gemini VIII Pilot's Report Press Conference in Houston Texas.

First Man   POST CONFORMED BLUE   62.

                          GEORGE MUELLER
                Neil, walk us through the decision
                to separate from the Agena.

As Neil considers, we **SMASH BACK TO** --

100   **INT. AUDITORIUM, MANNED SPACE CENTER – HOUSTON, TX – DAY**   100

Reporters **CLAMOR**. JULIAN SCHEER, the NASA HQ Assistant
Administrator for Public Affairs points to the Houston Post.

                      HOUSTON POST REPORTER
                You mentioned the rate of revolution
                was more than once a second. How
                near were you to being unconscious?

                          NEIL
                We didn't have any specific
                difficulty in observing the panel.

                      HOUSTON POST REPORTER
                You hadn't begun to gray out or
                anything like that?

Off Neil, HESITATING, not in his element, **SMASH BACK TO** --

101   **INT. CONFERENCE ROOM, MANNED SPACE CENTER – DAY**   101

                          GEORGE MUELLER
                Did you think to use the Agena to
                stabilize the combined craft?

                          NEIL
                We did. This was not successful.
                As I said, we initially assumed that
                the anomaly was with the Agena
                control system. There was no way to
                know a thruster on the Gemini was
                causing-- if we could've isolated
                each of Gemini thrusters, if we'd
                had that capability in the moment --

                                            SMASH BACK TO --

102   **INT. AUDITORIUM, MANNED SPACE CENTER – HOUSTON, TX – DAY**   102

Questions KEEP COMING. In **QUICK CUTS** --

                      AGENCE FRANCE REPORTER
                Agence France. Did you have any
                feeling of anxiety after the failure
                of the thrusters?

First Man   POST CONFORMED BLUE   63.

> HAMBURG PRESS
> In the midst of the spinning did you
> seem to realize or feel the presence
> of God closer than other times?

> TIMES REPORTER
> With this so hot on the heels of the
> loss of Charlie Bassett and Elliot
> See, do you question whether the
> program's worth the cost? In money
> and in lives?

SPIN BACK TO Neil. Blank, EXHAUSTED. **SMASH BACK TO --**

103 **INT. CONFERENCE ROOM, MANNED SPACE CENTER – DAY**   103

Neil and Dave still sitting there.

> GEORGE MUELLER
> Alright, thanks, guys. We've got a
> lot to discuss and we'll be back in
> touch with you soon.

It's not warm.

> NEIL
> Thank you.

As a hand **SHUTS OFF** the REEL-TO-REEL, we **CUT TO --**

104 **INT. NEIL'S OFFICE, ARMSTRONG HOUSE – DAY**   104

CLOSE ON Neil on the phone. As **AGITATED** as we've seen him.

> NEIL (INTO PHONE)
> *'Our Wild Ride in Space'*? It sounds
> sensationalist to me.

Neil stares at an ADVANCED COPY of a Life Magazine article.
We hear piano from the living room and we **INTERCUT WITH --**

105 **OMITTED**   105

A106 **INT. LIVING ROOM, ARMSTRONG HOUSE – SAME TIME**   A106

CLOSE ON small hands playing a piano. REVEAL Rick practicing.
Janet stands over him, trying to help.

> JANET
> We're using our thumb now, honey.

She hears Neil, inside his office.

First Man   POST CONFORMED BLUE   64.

IN HIS OFFICE, Neil walks back and forth, carrying the cradle.

> NEIL (INTO PHONE)
> Well, that's not my concern.

IN THE LIVING ROOM, Janet walks over to the table to wrap a birthday present from Mark.

> RICK
Seriously?
>   JANET
>   I don't want to hear it.

IN HIS OFFICE, Neil's frustration builds.

> NEIL (INTO PHONE)
> I'm not interested in how other
> magazines are framing the story, I
> think it's an inappropriate title
> for the piece.

Neil kicks the door shut.

IN THE LIVING ROOM, Rick stops playing, looks to Janet.

Janet reacts, sees Rick looking at her. She forces a smile, trying to pretend it's nothing.

IN HIS OFFICE, Neil puts down the cradle.

> NEIL (INTO PHONE)
> Well, then maybe you should take my
> name off it.

Neil hangs up, putting the receiver down with some **FORCE**. He walks back and forth in his office. <u>Not happy</u>.

He spots an old MODEL PLANE. Reacts. Looks off.

FLASH TO --

Karen. In Juniper Hills. Smiling at him.

FLASH BACK TO Neil. **STRUGGLING.** Off that look, **PRELAP --**

JIM: As is depicted here, the magazine initially titled this article 'Our Wild Ride in Space—By Neil and Dave.' Neil was extremely angry and put a stop to it by calling Hank Suydam, a writer assigned to Houston. Suydam wired *Life*'s editor-in-chief Edward K. Thompson who toned down the piece, took off the astronauts' byline and changed the title to 'High Tension Over the Astronauts.' Of course, *Life* ran another article in its next issue entitled, 'Wild Spin in a Sky Gone Berserk.' And while Neil and Dave got to write their own piece—'A Case of Constructive Alarm'—it was so heavily edited that Neil again complained.

JOSH: We do take some license here. Dave Scott was present for this call, which actually happened at work. But we wanted to understand the emotional impact on Janet and the family, so we elected to play the call at home. Also, Neil was known to be pretty even tempered, but Ryan pushes it a little to convey just how angry he was—and how hard it was to be in the public eye for these missions.

LEFT: Neil and Dave are questioned in the formal mission review as backup Gemini VIII commander Pete Conrad (Ethan Embry) looks on.
RIGHT: Deke and Gilruth listen to a response from Neil.
BOTTOM: Gilruth listens carefully as NASA Associate Administrator George Mueller (Kermit Rolison) considers a response from Dave.

First Man   POST CONFORMED BLUE   65.

> PAT **(PRELAP)**
> If it's any consolation, Ed was a
> zombie for weeks after Gemini Four.

106   **EXT. ED & PAT WHITE'S HOUSE - LATE AFTERNOON**   106

Janet stands with Pat beside a lemonade stand Bonnie set up.

> JANET
> Yeah?

> PAT
> Uh huh.

> JANET
> Yeah, I guess it must be...
> disorienting for them.

Pat nods. A beat.

> JANET
> God, I married Neil because I wanted
> a normal life.

Pat chuckles. Janet laughs too.

> JANET
> I know. He was just so different
> from all the other boys on campus.
> He'd been through the war, you know.
> He knew what he wanted to do. It
> seemed so stable.
>   (then)
> I guess all I wanted was stability.

Janet's smile fades. A darkness crosses. Pat reads it.

> PAT
> I've got a sorority sister with a
> normal life.

> JANET
> Yeah?

> PAT
> She married a dentist.

> JANET
> A dentist. Sounds good.

> PAT
> He's home by six every night. And
> every few months she calls to say
> she wishes he weren't.

Janet manages a smile. A beat, **CUT TO --**

"BOTH MARY AND NATHAN CREATED AN AUTHENTIC
PALETTE FOR THE ACTORS, DAMIEN, AND LINUS
TO DO THEIR WORK IN."

WYCK GODFREY, PRODUCER

First Man   POST CONFORMED BLUE   66.

107   **INT. NEIL'S OFFICE, ARMSTRONG HOUSE – LATE AFTERNOON**   107

A LARGE DESK.   **SCHEMATICS** of the Agena, **DIAGRAMS** of the Gemini
thrusters... <u>Neil hunched over it all</u>.   We hear a **KNOCK**.

>   NEIL
>   Yeah.

Ed walks in.

>   ED
>   Still working, I see?

>   NEIL
>   Yeah.

Ed clocks the copy of LIFE in the waste basket.

>   ED
>   Well, I was, uh, gonna go grab a
>   beer at Dave's.

No response from Neil.   Ed, giving up, starts to retreat...

>   ED
>   Alright...

>   NEIL
>   ...I could use a beer.

Ed pauses.   He and Neil share a smile.   **PRELAP** MUSIC, **CUT TO** --

108   **EXT. BACKYARD, DAVE SCOTT'S HOUSE – NIGHT**   108

Neil and Ed drink with Dave on the patio; through the window,
we see the shadow of Dave's wife doing dishes.

The three men listen to the **RADIO**.

>   DAVE SCOTT
>   You know, I will say one thing.
>   It's all I can think about. Getting
>   back up there.

Dave looks down.

>   ED
>   You just caught a rough break. You
>   boys did everything right. I was
>   talking about it with Gus and we
>   both agreed.

He says it to both of them, but it's directed at Neil.

JOSH: This scene is a dramatization, but we imagined these three guys probably had a beer at some point. Of course, Ed and Gus were selected for the first Apollo crew in January 1966—we play with time here so we can show Neil's reaction to the news.

JIM: It's an interesting "what if," the bit about Gus being first to land.

JOSH: That comes from Deke Slayton's autobiography *Deke!*. Deke says that had Gus not been killed in 1967, he would've been first to land.

JIM: I know Deke said that, but missions were assigned years in advance and back in 1966 no one knew which mission would be the landing. So while Deke might have arranged

things to give Gus a great chance at landing, given the care with which Deke selected his commanders and lined up his crews, if the greater mission schedule changed I don't think Deke would have changed things up just so Gus could be first.

JOSH: No. But all we're really saying here is that Deke had a strong desire to have Gus (a fellow Mercury astronaut and a good friend) be first. So any guy lucky enough to be part of Gus's crew had a good shot at winning the lunar landing lottery.

There's also a larger point—Neil wasn't ordained to be first. He was qualified, maybe most qualified to make the landing. But, as I believe Neil himself would have pointed out, some of this was happenstance. At the very least, if Gus hadn't died, Neil might have been bumped down the rotation and Apollo 1 (or another Apollo mission) might have been first instead.

ABOVE: The real Apollo1 crew: Ed White, Gus Grissom, and Roger Chaffee.

First Man   POST CONFORMED BLUE   67.

                    DAVE SCOTT
        Yeah, I heard a rumor you'd been
        hanging out with Gus.

                    ED
        You did?

                    DAVE SCOTT
        Ed... is it true?

Neil looks at Ed.

                    NEIL
        Is what true?

Ed hesitates. Can't help but smile.

                    ED
        Deke pulled me aside and told me he
        and Gus want me on the crew.

                    NEIL
        For the first Apollo?

Ed nods. Dave explodes, excited for Ed. Neil GRINS.

              DAVE SCOTT
Yeah, yeah. Holy shit.   NEIL
That's huge!   Congratulations. I've gotta
                                shake your hand.

Neil smiles, stands, holding out his hand. Ed smiles, shakes.

                    ED
        Thanks, man.

                  NEIL
Saturn's a monster.   ED
                      It is.

                    NEIL
        You're in for one heck of a ride.

                *DAVE SCOTT
        And hey! You know Deke wants Gus to
        be the first one on the Moon, so...
        this puts you in the LM with him for
        the landing.

                  ED
Let's not get carried away   DAVE SCOTT
here...   Come on, Ed...

But Ed's not having it. Dave shakes his head, playful.

JIM: Two weeks after the flight, the Gemini VIII Mission Evaluation Team ruled out pilot error as a factor in the emergency. In revealing the team's findings, Bob Gilruth declared that the crew, in fact, demonstrated remarkable piloting skill in overcoming the problem. There was no question that Armstrong would be given another assignment as a mission commander.

JOSH: It was also clear that in the changing political climate of the late 60s, had Neil and Dave not overcome the problem, it might have jeopardized the future of Apollo. We have Deke use Dave Scott's word for that— "showstopper."

"HE'S DONE A LOT OF THINGS IN HIS LIFE WHERE HE WAS MR. COOL WHEN THE WORLD WAS GOING CRAZY AROUND HIM AND HE SAW HIS WAY THROUGH. GEMINI VIII IS A PERFECT EXAMPLE. I DON'T KNOW OF ANYBODY WHO COULD HANDLE GEMINI VIII THE WAY HE DID."

ALFRED WORDEN, COMMAND MODULE PILOT
FOR THE APOLLO 15

First Man   POST CONFORMED BLUE   68.

                         DAVE SCOTT
               Alright, get out of my house, I'm
               gonna go to bed. I'm not kidding,
               get out of my house.

The men laugh and we **CUT TO** --

110   **INT. GILRUTH'S SECRETARY'S OFFICE, MSC - DAY**   110

Neil, in a chair, in the reception area. As anxious as he
gets. SWEATING a little. At last, the door opens.

                         DEKE
               Neil.

Neil stands. Quickly. Deke leads him into --

111   **INT. GILRUTH'S OFFICE, MANNED SPACE CENTER - DAY**   111

Wood paneling, a few service plaques, function over form.
Gilruth sits behind a desk.

                         GILRUTH
               Hey, Neil. Don't bother sitting,
               it's gonna be a short meeting.

Neil stands. A tense beat.

                         GILRUTH
               We've talked it through and we think
               it's pretty clear. If you hadn't
               kept cool, well, you wouldn't be
               here and we'd still be asking what
               the hell happened.

                         *DEKE
               So would Congress. It's a
               showstopper.

                         GILRUTH
               This mission was a success. We're
               full steam ahead for Apollo. You
               good with that?

                         NEIL
               ...yes sir.

It's a BIG MOMENT. Even Neil can't mask his **RELIEF**.

                         GILRUTH
               I trust you won't mind representing
               us at the White House?

First Man   POST CONFORMED BLUE   69.

      NEIL
  No sir.

      GILRUTH
  Good.

Off Neil, we **CUT TO --**

112 **EL LAGO (APOLLONIA) MONTAGE**   112

In a series of QUICK CUTS, we see --

112A ARMSTRONG KITCHEN. Neil studies a preliminary Apollo flight112A
plan... at least until he realizes that someone's put his
wallet on his head. He smiles, grabs Rick.

112B ARMSTRONG HALL/BEDROOM. Neil creeps down the hall, in pursu112B
of the boys. He slips into the bedroom; Rick runs out but
Neil finds Mark, playfully throws him up over his shoulder.

112C ARMSTRONG KITCHEN. Neil tries to punish Mark, makes him stand12C
in the corner. Janet covers a smile, then can't help it,
starts laughing. Neil can't help but laugh too...

113 ARMSTRONG POOL. Summer chaos. Green trees, bright sun. 113
Local kids in the pool. Rick dives in as the Whites' dog
barks at Eddie White, using a HOPPITY HOP to splash Bonnie and
Carrie, floating in the USS Eggshell. Off the joy, **PRELAP --**

      ED **(PRELAP)**
  I tell you Eddie started asking
  questions about the new command
  module?

109 **EXT. EL LAGO STREET - LATER THAT NIGHT**   109

Ed and Neil walk home.

      NEIL
  Is that right?

      ED
  Yeah, he wants to, wants to know if
  it's gonna fly any different from
  Gemini. If all the buttons are
  gonna be in the same place.

Neil smiles, looks over at Ed.

      NEIL
  Oh boy. You've got yourself a
  little engineer there.

Ed laughs, looks up at the Moon.

"MY WORK REQUIRED AN ENORMOUS AMOUNT OF MY TIME. I DIDN'T GET TO SPEND THE TIME I WOULD HAVE LIKED WITH MY FAMILY AS THEY WERE GROWING UP."

NEIL ARMSTRONG

JOSH: I love every bit of the Apollonia Montage. It's some of my favorite stuff in the movie. This, despite the fact that not a bit of the dialogue is scripted. Everything in here is ad-libbed, just Damien playing with the actors. That's why it's so real.

We originally had a scripted scene set at the pool, but it didn't shoot that well. And frankly, even if it had I'm not sure if it would have achieved what Damien and I wanted—a last idyllic moment of home life for Neil before Ed dies, and before Neil becomes solely focused on the mission. To me, this says a lot about screenwriting—you're just drawing a blueprint. Sometimes what's on the page works, and sometimes it's transformed into something totally different, something that gets across the intention of the scene much more elegantly than the dialogue you wrote. All of which is to say while I believe films need a singular vision, I also believe that the best version of any film utilizes the various talents of all the people assembled to make it. It's truly a collaborative process.

JIM: I like that the script has Neil open up and talk about Karen with Ed in Scene 109. But I wonder whether this would have happened. Most people who knew Neil said he never brought up Karen's illness; several of his closest colleagues didn't even know he'd had a daughter. And Janet said Neil rarely talked about Karen. Emotionally, Neil was a tightly packaged man.

JOSH: Absolutely. We're just trying to show how Neil and Ed were close. We have Ed mention his faith, something Ed's daughter Bonnie stressed in our conversations with her. And we have Neil bring up Karen, which helps emphasize how challenging we thought this immediate post-mission moment would have been for him. By the way, you'll note that Scene 109 comes after Scene 113. Scene 109 was originally set after the beer at Dave's—we moved it early in post. I've left the original scene numbers intact so readers can see how fluid the storytelling can be in editorial.

First Man POST CONFORMED BLUE 70.

> ED (CONT'D)
> I tell you, though, I love that he's
> interested. He came in the other
> morning, he comes running up and he
> says, "Daddy, if you go to the Moon,
> are you gonna be lonely out there?
> So far away from earth? All of us
> back here at home?"
> (then)
> This whole thing is expanding his
> horizons... It, uh, you know, it
> gives me faith. Make sense?

But Neil's distracted. He eyes a **SWING** in a neighbor's yard.
Ed notices him staring.

> NEIL
> Wittry's got a new swing set.

> ED
> Yeah, I noticed that.

Neil walks on for a beat.

> NEIL
> We had a swing like that back up in
> Juniper Hills.
> (then)
> Karen really loved it.

> ED
> ...that's your daughter.

Neil doesn't answer. He looks again at the swing. Ed starts
to say something but...

> NEIL
> I guess I oughta be getting home.

Neil goes. Off Ed, realizing how deep that wound is --

114 **OMITTED** 114

A115 **OMITTED** A115

116 **INT. GANTRY ELEVATOR, LAUNCH TOWER, PAD 34, KSC - DAY** 116

Deke rides the elevator with Gus, Ed and Chaffee, spacesuits
and helmets with visors up. Ed, all **NERVOUS EXCITEMENT**,
wouldn't trade this for the world. But Gus is concerned.

> GUS
> I think the 21st is pushing it.

"THIS IS A VERY GRAND STORY, BUT IT'S ALSO A VERY INTIMATE, PERSONAL STORY. WE DECIDED TO SHOOT ON SIXTEEN MILLIMETER, WHICH IS GRAINIER AND FEELS MORE POETIC."

LINUS SANDGREN, CINEMATOGRAPHER

JOSH: On January 27, 1967, Neil, Gordon Cooper, Dick Gordon, Jim Lovell, and Scott Carpenter were in Washington for the signing of the Treaty on Principles Governing the Activities of States in the Exploration and Use of Outer Space. Pete Conrad wasn't there, but he's a larger than life character and we wanted to keep him present in the movie, so we included him in this sequence.

JIM: The treaty, still in effect today, precluded land claims on the Moon, Mars, or any other heavenly body. It also outlawed the militarization of space. Following the signing, there was a reception in the Green Room of the White House, hosted by President Johnson and his wife. Many political heavyweights were there, including Senators Everett Dirksen, Eugene McCarthy, Albert Gore Sr., and Walter Mondale. I like the dialogue here.

JOSH: Damien and I were fascinated by how the astronauts could remain so focused even as the political climate changed so dramatically around them.

JIM: Most people believe the U.S. space program enjoyed broad support during the Space Race, but public opinion in favor of a Moon landing rarely rose above 50%, and the percentage of people who thought we were spending too much on space climbed over the course of the decade.

JOSH: We wanted to highlight that here.

"WE HAVE TO PREFIT EVERYBODY. EVERY BACKGROUND, EVERY SPEAKING PART. EVERYBODY GETS FIT IN THIS MOVIE."

MARY ZOPHRES,
COSTUME DESIGNER

ABOVE: Gordon Cooper (William Lee), Pete Conrad (Ethan Embry) and Jim Lovell (Pablo Schreiber) pose with a number of dignitaries.
BELOW: Damien's parents made a cameo in this scene. And his sister, Anna, plays the White House Staffer in Scene A121.

>   * DEKE
> The Russians have already
> tested the Soyuz.

           GUS
       I'm aware the Russians've
       already tested it, but if the
       command module's not ready --

      ED
It'll be ready, Gus.

>   *GUS
> I'm not going up there in a goddamn
> lemon.

      DEKE
No, we wouldn't let you. If the
ship doesn't pass plugs out, we'll
go back to the drawing board.

The elevator stops. **Short.**   <u>A FOOT below the bridge to the</u>
<u>**APOLLO COMMAND MODULE**</u>. Gus **SHOOTS** Deke a look, we **CUT TO** --

115   **INT. GREEN ROOM, THE WHITE HOUSE - DAY**   115

Lovell and Neil, in SUIT AND TIE, talk to a SENATOR. Neil
clings to a wine glass. Around him, CHATTER and PIANO.

      LOVELL
Well, we're very, very bullish on
Apollo, Senator.

>   *SENATOR
> I should hope so, given the time
> we've spent developing it. Times
> have changed, you know. Half the
> country doesn't think it's worth it
> anymore.

      NEIL
We only learned to fly sixty years
ago, so I think if you consider the
technological developments in the
context of history, it's really not--

      SENATOR
I'm considering it in the context of
taxpayer dollars.

      LOVELL
    (stepping in)
And so are we, Senator. Between us,
we're doing some final tests on the
new command module today.
      (MORE)

First Man   POST CONFORMED BLUE   72.

> LOVELL (CONT'D)
> I'm sure Mr. Gilruth would be happy
> to tell you about it. Let me
> introduce you to Bob, come on...

Lovell leads him away. Neil eyes the UNCTUOUS POLITICOS, Pete
Conrad working on a few of them. Off Neil, maybe wondering
how long until he can get back to work, we **CUT TO --**

117   <u>**INT. THE WHITE ROOM, ADJUSTABLE LEVEL 8, PAD 34 – MOMENTS LATER**</u>17

The Pad Leader oversees techs removing thick power cords from
the new APOLLO COMMAND MODULE.

> PAD LEADER
> Closing hatches now.

Ed, in the capsule, helmet on, pushes the inner hatch towards
us, <u>SEALING THE COMMAND MODULE.</u>

Techs close the hatch above, using one RATCHET to tighten all
SIX LATCHES... They close the cover, <u>start the process again.</u>

> **INTERCUT WITH --**

118   <u>**INT. APOLLO COCKPIT, PAD 34 – SAME TIME**</u>   118

CLOSE ON another a hand using a ratchet to lock the hatch.

> ED (O.C.)
> Okay, that's all of 'em.

FIND Ed in the familiar Apollo cockpit. He glances at the
locked latches, puts away his ratchet.

<u>**IN THE WHITE ROOM.**</u>   The techs close up the ablative hatch.

> PAD LEADER (COMMS)
> *Ablative hatch closed. Closing the*
> *boost protective cover.*

They close the boost protective cover, then start to remove
all of the hoses and wires connecting the craft to the ground.

> PAD LEADER (COMMS)
> *And plugs out.*

The techs finish up, leave the White Room.

<u>**IN THE APOLLO COCKPIT.**</u> The men sit, ready to start the test.

> GUS (INTO COMMS)
> Ready for oxygen purge.

JOSH: While Washington's priorities might have changed during the 60s, NASA remained focused on the Space Race.

JIM: The Russians had tested the Soyuz in November 1966 and were gearing up to test the Proton-K rocket (they would test it in March 1967 and send it round the Moon the following year).

JOSH: While at this moment it looks as if Scene 116 may end on the cutting room floor, we highlight the Soviet's progress there to give additional context to the Apollo 1 fire—which occurred on January 27, 1967, while Neil was at the White House.

JIM: The fire was especially shocking as it came not during a mission, but during a routine test of the new Apollo Command Service Module (CSM).

JOSH: We try to make that clear. We also wanted to highlight the concerns the crew had about the new CSM— they'd gone so far as to present Joseph Shea (the CSM design/construction manager) with a portrait of themselves, hands clasped in prayer, around a model of the spacecraft.

Various issues were discovered in testing and NASA had to make so many changes to the CSM that the training simulators couldn't keep up with them. Which led Gus to hang a lemon on the CSM simulator. This inspired our lemon line.

JIM: Right. But he didn't call the actual CSM a lemon.

JOSH: No, but given the number of concerns the astronauts had about the CSM, we imagined he might have. According to Jim Lovell, backup commander Wally Schirra warned Gus and Shea about the craft after the final altitude chamber test: "There's nothing wrong with the ship that I can point to, but it just makes me uncomfortable. Something about it doesn't ring right."

BELOW: Gus Grissom (Shea Whigham), Ed White (Jason Clarke), and Roger Chaffee (Cory Smith) in the Apollo 1 Command Module for the plugs out test.

We hear the **HISS** of oxygen... and **SNIPPETS** of conversation over Comms. Gus, beside Ed, hits the COMMS BUTTON.

> GUS (INTO COMMS)
> Guys, you want to hold down the chatter? We've got an open mic.

> ROCCO PETRONE (COMMS)
> *Uh, let's hold the countdown.*

A beat. The crew looks annoyed as they wait for further instructions from LAUNCH DIRECTOR **ROCCO PETRONE.**

> ROCCO PETRONE (COMMS)
> *Sorry guys, we'll get this squared.*

> GUS (INTO COMMS)
> Shit, we're gonna be here all
> night.

> ROCCO PETRONE (COMMS)
> *...Gus, we didn't get that.*

> GUS (INTO COMMS)
> Course you didn't.
>   (off Ed's laugh)
> Glad you think this is funny.

Ed smiles. Gus, PISSED, shakes his head, we **TIME CUT TO --**

119 **INT. APOLLO COCKPIT, PAD 34 - FOUR HOURS LATER (EVENING)**   119

It's dark out the window now. Ed reads a manual, BORED. Gus, frustrated, sweating, lifts up his visor, rubs his face. **HOLD** for <u>a long beat</u>. Then Roger gives it a shot.

> CHAFFEE (INTO COMMS)
> Well I haven't talked yet, how's this? 1, 2, 3, 4, 5, 4, 3, 2, 1.

No response. **UNNERVING.** Unsettling. **STATIC** over comms.

> DEKE (COMMS)
> *...ah, we need another minute to get it sorted.*

> GUS (INTO COMMS)
> How are we gonna get to the moon if we can't talk between three buildings?

No response. It's warm. Ed sweats, lifts his visor too.

> ED (INTO COMMS)
> They can't hear a thing you're saying.

> GUS (INTO COMMS)
> Jesus Christ.

> ROCCO PETRONE (COMMS)
> *Say again?*

First Man   POST CONFORMED BLUE   74.

>   GUS (INTO COMMS)
>   (annoyed)
>   I said how are we gonna get to the
>   Moon if we can't talk between two or
>   three buildings?

Nothing. Ed chuckles.

>   ED (INTO COMMS)   GUS (INTO COMMS)
>   You tell 'em, Gus.   Mickey Mouse shit.

>   TECH (COMMS)   ROCCO PETRONE (COMMS)
>   *I got a surge in the AC Bus 2*   *Try resetting the meter.*
>   *Voltage.*

>   DEKE (COMMS)
>   *You getting this, Gus?*

Gus, happy to have something to do, leans forward.

>   GUS (INTO COMMS)
>   Rog, you pick anything up on the
>   dials?

Now Chaffee looks... and spots a spark... then a **SMALL FLAME**
on the floor. Before he can react, the flame **JUMPS**.

>   ED (INTO COMMS)
>   Hey, we got a fire in the cockpit.

Flames **FLASH** through the cockpit, **FIRE BLAZING** around them.

Ed grabs the ratchet, loosens the latches and pulls at the
hatch... *but it won't budge, the pressure inside too great.*

>   CHAFFEE (INTO COMMS)
>   We got a bad fire! We're burning up in
>   here!

ON ED, REALIZING, as FIRE **RUSHES IN** with speed as shocking as
it is terrifying and THE **FLASH** OF AN **EXPLOSION** SMASHES US TO --

A120 **INT. WHITE ROOM, PAD 34, KSC - SAME TIME**   A120

CRACK! The boost protective cover **RUPTURES**, smoke trickling
out and we **SMASH TO** --

120 **OMITTED**   120

A121 **INT. GREEN ROOM, THE WHITE HOUSE - EVENING**   A121

Neil's at the bar. A WHITE HOUSE STAFFER walks up.

IM: As depicted, this was a long day for the crew.
They entered the hatch at 1 p.m. and had several issues
that led to held countdowns. The first was a strange odor Gus
noticed and the second was the problem with the comm loop.
This problem actually happened a bit later in the day than
depicted—the count was held at 5.40 p.m. due to this issue.

OSH: The comms in Scene 119 are from the transcripts and
the audio, which is now public on YouTube. We compress
time a bit, as the Chaffee line at the top of the scene happened
about 6.27 p.m.—three and a half minutes before the fire (6.31
p.m.). We also add some dialogue around the AC Bus 2 voltage
surge, which happened nine seconds before the fire. But the
rest of what's here is pretty much verbatim. It was important
to us to be as true as possible to what was said by these guys
in their final moments.

IM: Of course, the transcript itself is not without
controversy. Acoustical experts brought in by NASA to listen
to these recordings aren't in total agreement about which
astronaut was saying what. But what you've got here is the
general consensus.

OSH: One visual note—in reality, the entire White Room
filled up with black smoke when the command module
ruptured. We chose to limit the smoke as we thought it was
a more effective way to keep the viewer focused on what
happened to the men inside the Command Module.

JOSH: In reality, the astronauts did not learn about the Apollo 1 fire until they got back to their hotel, at 7.15 p.m. that night.

JIM: When Neil got to his hotel room, his message light was on. He had an urgent message from the Apollo program office. He called and an anonymous man told him what had happened.

JOSH: We had Neil call Deke to emphasize Deke's role in the program. And the scene was originally written in Neil's hotel room, but it was easier for production to set it outside our Green Room. In every production, no matter how big the budget, you make some sacrifices. This was a relatively small one, as the spirit of what happened remains the same.

JIM: The breaking of the glass is dramatic license.

JOSH: Yes, as you mentioned, emotionally, Neil was tightly packaged. Even in this intense moment of tragedy, I'm not sure Neil would have displayed that much emotion. But we wanted to convey how devastating this fire was, both for the program and Neil personally; that even someone as controlled as Neil might not have been able to keep his emotions in check.

JIM: An early draft had him repeatedly slamming down the phone. I think I pushed back on that.

JOSH: You did, which was helpful. Breaking the glass was a more subtle way to convey the same idea, and more in keeping with who Neil was. Again, it's likely that Neil didn't outwardly react at all. But we felt we needed some demonstration of the excruciating pain of this tragedy.

First Man   POST CONFORMED BLUE   75.

                         WHITE HOUSE STAFFER
                 Mr. Armstrong? I have Deke Slayton
                 on the phone for you.

Neil's a bit surprised, but he's happy for the distraction.
He takes his glass as the staffer leads him out into --

121   **INT. HALLWAY, THE WHITE HOUSE – CONTINUOUS**   121

A black phone on small side table. Neil picks up.

                         NEIL
                 Thank you.
                   (into phone)
                 I'm glad you called, I'm not sure if
                 I'm helping or hurting over here.

          DEKE (OVER PHONE)
Neil, we had a problem with
the plugs out test.   NEIL (INTO PHONE)
                                       That's why we have tests,
                                       right? We'll figure it out.

               DEKE (OVER PHONE)
          ...there was a fire. There's no
          easy way to say this... Ed, Gus and
          Roger, they're gone.

Wait... *what?* Neil tries to process. His eyes **DARKEN.**

                    DEKE (OVER PHONE)
              Neil, listen, we need you guys to
              head back to the hotel. The press
              is going to be all over this,
              Congress is going to be calling for
              investigations, we just don't want
              you guys in the middle of all that.
                (then)
              Do you understand?

                         NEIL (INTO PHONE)
                 Yeah.

          DEKE (OVER PHONE)
Alright then.   NEIL (INTO PHONE)
                                       Okay.

Neil slowly hangs up. He sits, his face TAUT. He starts to
SHAKE, CLENCHING his jaw...

Until we hear a **POP.**

Neil FLINCHES. Sees his hand is <u>covered in</u> **BLOOD**, the glass
he was holding **IN PIECES.**

JIM: After Neil and the other astronauts heard about the Apollo 1 fire, they assembled in a suite at the Georgetown Inn. As discussed, Conrad wasn't there. But they did share a bottle of scotch and talk 'til the wee hours.

JOSH: CBS coverage of the tragedy can be found on YouTube. It's haunting; CBS rebroadcast interviews with all three crew members. We had Jason recreate the portion of the interview with Ed that CBS aired that evening. It's heartbreaking.

JIM: I can only imagine how Neil must have felt. He'd lost Karen, Elliot, and Ed, almost died in Gemini VIII... you don't even mention the fact that Joe Walker, Neil's close friend and mentor from Edwards, was also killed in a freak plane crash in June 1966.

JOSH: The grace and endurance Neil showed in the face of all this loss seemed truly heroic to us.

A beat. Neil reaches into his coat pocket for a handkerchief and wraps the wound. His face blank...

...save for his eyes. <u>He is quietly devastated</u>.

As he stands there, still processing, we **PRELAP** --

> ED (**PRELAP**, ON TV)
> *I think a lot of people forget about the influence that the lunar program has on the raising of our young people in the country.*

122   **INT. CONRAD'S ROOM, GEORGETOWN INN - LATER**   122

CLOSE ON a TV. Footage of a CBS interview. With Ed.

> ED (ON TV)
> *I think that if a civilization doesn't look out, if it doesn't try to expand its horizons, then we're not going to progress as a nation.*

Neil, Conrad and Lovell now in shirtsleeves, stare at the screen. And drink. CBS cuts to anchor MIKE WALLACE.

> MIKE WALLACE (ON TV)
> *At 10:30 tonight Eastern Time, rescue teams began to remove the bodies of the three astronauts from the charred spacecraft. A NASA spokesman said the dead astronauts were left in the ship for four hours to aid the investigation into the tragedy.*

PUSH IN on Neil. His eyes are full of **PAIN. ANGER.** But as Wallace talks, we see him slowly **PUSH THOSE EMOTIONS DOWN**...

> MIKE WALLACE (ON TV)
> *And according to the latest information from NASA at the Manned Spacecraft Center in Houston, the first Apollo flight, which was scheduled for February the 21st, has now been postponed indefinitely...*

...and as Neil's eyes **HARDEN**, we...

                                   FADE TO BLACK.

First Man   POST CONFORMED BLUE   77.

Ellington Air Force Base
1968

SMASH TO --

123 **EXT. ELLINGTON AIR FORCE BASE - HOUSTON, TX - DAY** 123

TIGHT ON NEIL'S FACE. Eyes focused. *Wounds buried, scar tissue invisible to most, less so to us.*

> FRANK BORMAN (COMMS)
> *Winds are pretty rough today, keep*
> *an eye on your yaw.*

PULL BACK to find Neil exposed to the elements, strapped into **THE LUNAR LANDING TRAINING VEHICLE**...

A MESS of METAL PIPES with a COCKPIT.

Neil hits the thruster. A BURST of peroxide **JOLTS** the craft left and we **PULL BACK FURTHER** to FIND Neil... **1000 FEET IN THE AIR** in a contraption that *doesn't look like it should fly.*

Jesus. The camera does a **WILD 360** around the belching craft then... **PUSHES IN** on Neil.

> NEIL (INTO COMMS)
> One thousand feet. Switching to
> lunar mode and starting descent.

Neil switches from gimbal lock to lunar simulation mode. He begins to guide the LLTV to a landing.

Thrusters **POP** and **HISS**, the jet engine below him **ROARS** and the craft **BUCKS**, but Neil's eyes REMAIN STEADY, ticking from his surroundings to his gauges.

CLOSE ON the ALTITUDE GAUGE: *700... 400... 200 feet.*

> NEIL (INTO COMMS)
> Final landing approach.

Neil pilots the LLTV down through 100 feet... when the bottom **FALLS OUT.** The craft starts to **DROP RAPIDLY.** Neil **PULLS** on the thruster, but the LLTV doesn't respond.

> FRANK BORMAN (COMMS)
> *You're too low. Neil! Climb!*

Neil **SWITCHES** back to GIMBAL LOCK...

The craft **SHOOTS UP** to 200 feet... Neil tries to regain control, but the craft **QUICKLY SLIDES RIGHT!!**

Thrusters **SPIT** steam; the peroxide burns Neil's neck as he tries to correct the roll... but the craft goes ***UP ON ITS SIDE LIKE A CARNIVAL RIDE!!***

> FRANK BORMAN (COMMS)
> *Neil, slow your rates! Neil, do you read?*

> NEIL (INTO COMMS)
> Control is degrading.

But Borman's voice **FADES**... then **THE SOUND DROPS OUT ENTIRELY.** In the **SILENCE** we **PUSH IN TIGHT** on Neil's face. We clock the **INTENSITY, HARD** and **COLD,** the **DARKNESS** in his eyes...

...MORE FRIGHTENING than anything happening around him. Neil's eyes **TICK DOWN,** the ground **RUSHING UP** to meet him.

LIGHTENING FAST Neil PULLS the ejection handle. **BOOM!!!** Sound comes **RUSHING** back as the Ejection Seat **EXPLODES** out of the LLTV, **TOSSING** Neil up into the air!!!

POP! Neil's parachute unfurls, stabilizing him just in time to see the LLTV CRASH below... and **BURST INTO FLAMES.**

Neil floats down quickly, lands **HARD**... and is immediately **YANKED** back by his parachute... which **DRAGS** him through the high grass. Neil **STRUGGLES** to stop rolling...

...at last coming to a halt. He STAGGERS to his feet, face red, bleeding... the LLTV a **FIERY BLAZE OF METAL** in the b/g.

> DEKE **(PRELAP)**
> The vehicle is not safe.

124 **INT. WORK SPACE, MANNED SPACE CENTER – DAY**   124

A huge hangar, techs look over F-1 ROCKET ENGINES. Gilruth and Deke walk with Neil, bad SCRATCHES on his face and neck.

> NEIL
> Unfortunately, it's the best simulation we have.

Neil is fairly calm; Gilruth and Deke less so.

> GILRUTH
> You and the others are too valuable.

> DEKE
> It's a fly by wire system, it's got no backup.

> NEIL
> The ejection seat is the backup.

JIM: Neil had started working on a lunar landing simulator back when he was at Edwards. The basic idea was to mount a jet engine underneath the vehicle; the jet would lift the vehicle to a test altitude, then the pilot would throttle back the engine to support ⅚ of the vehicle's weight (simulating the Moon's ⅙ gravity). Two lift rockets would control rate of descent and horizontal movement; several smaller thrusters would give the pilot further attitude control. Neil's old boss and good friend Joe Walker first tested the Lunar Landing Research Vehicle (LLRV) in 1964. In 1966, NASA then decided to turn the machine into a full-fledged trainer.

JOSH: Damien and I were immediately captivated by the Lunar Landing Training Vehicle (LLTV), which the astronauts liked to call the "flying bedstead." It was like nothing we had ever seen. And it exemplified the dangerous lengths to which these men went in training for the lunar mission.

JIM: Frank Borman called flying it "a hairy deal." Jim Lovell said the counterintuitive nature of the controls and the different safety factors "made me worry about flying it." And Buzz Aldrin noted, "Without wings, it could not glide to a safe landing if the main engine or the thrusters failed. And to train on it properly, an astronaut had to fly at altitudes up to 500 feet. At that height a glitch could be fatal."

JOSH: Neil saw that first hand on May 6, 1968. You can find footage of this crash on YouTube; it's incredible.

JIM: Neil's description is as powerful as the footage. "I wouldn't call it routine, because nothing with an LLTV was routine, but I was making typical landing trajectories... and in the final 100 feet of descent going into landing, I noted that my control was degrading. Quickly control was nonexistent. The vehicle began to turn. We had... no emergency system with which we could recover control. So it became obvious as the aircraft reached 30° of banking that I wasn't going to be able to stop it. I had very limited time left to escape the vehicle... The ejection was somewhere over fifty feet of altitude, pretty low..."

JOSH: Those who observed the accident felt Neil was lucky to be alive. According to Chris Kraft, if Neil had tried to eject as much as ⅖ of a second later he would have died.

JIM: Christian Gelzer, the Chief Historian at NASA's Armstrong Flight Research Center (and my former student) was a huge help here.

JOSH: Absolutely. There was no published transcript of this flight so Christian helped us with the dialogue; he also trained Ryan and the rest of us on the LLTV so we could try to replicate Neil's work in the cockpit.

ABOVE: The real Lunar Landing Training Vehicle (LLTV).

LEFT: A number of different means were used to replicate the LLTV. The cockpit was attached to a gimbal rig in front of LED screens on stage. The cockpit was also shot outdoors on the gimbal rig with Linus Sandgren manning the camera on its side. In addition, the team also used a crane and to shoot the mock trainer.

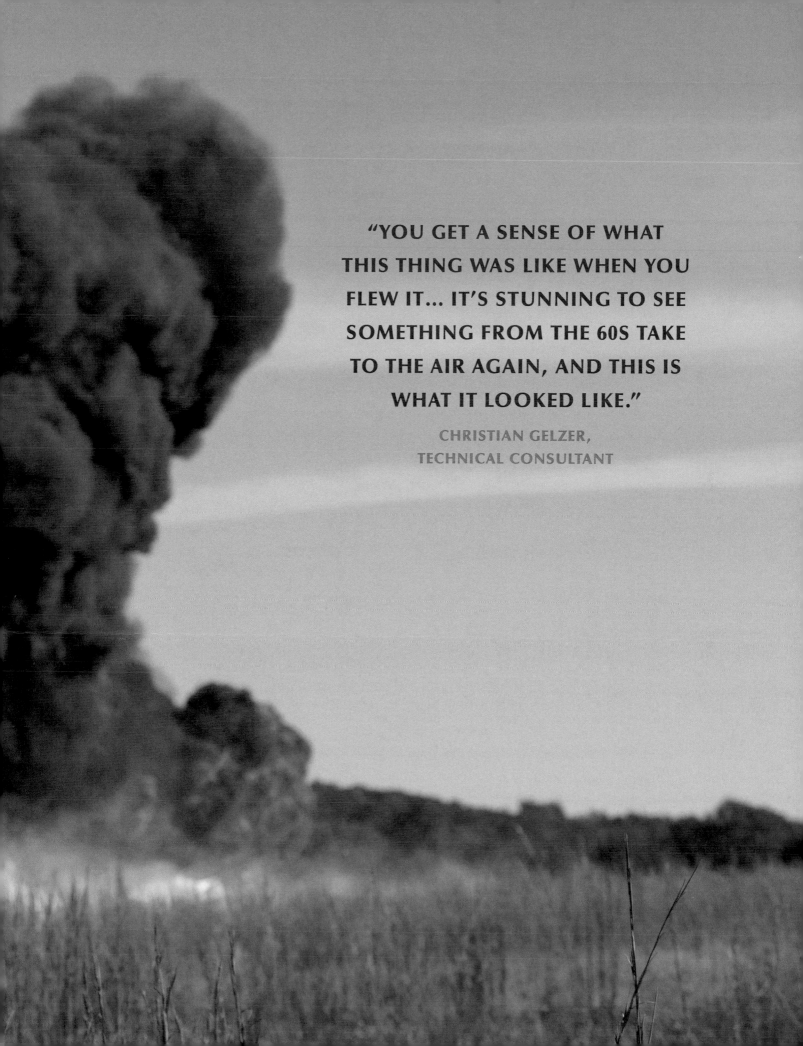

"YOU GET A SENSE OF WHAT
THIS THING WAS LIKE WHEN YOU
FLEW IT... IT'S STUNNING TO SEE
SOMETHING FROM THE 60S TAKE
TO THE AIR AGAIN, AND THIS IS
WHAT IT LOOKED LIKE."

CHRISTIAN GELZER,
TECHNICAL CONSULTANT

First Man   POST CONFORMED BLUE   79.

> GILRUTH
> Neil, the political fallout
> from another accident --

>> NEIL
>> With all due respect, it's
>> not my job to worry about the
>> political fallout.

> DEKE
> The damn thing could have
> killed you.

>> NEIL
>> Well, it didn't.

> DEKE
> A split second more and --

>> NEIL
>> We need to fail.

Neil stops, turns to face them.

> *NEIL
> We need to fail down here so we
> don't fail up there.

> GILRUTH
> Neil, at what cost? Huh?

> NEIL
> At what cost? It's a little late
> for that question, isn't it sir?

Neil turns, heads off. Off Gilruth and Deke --

A125   **INT. ARMSTRONG HOUSE - LATE AFTERNOON**   A125

Janet walks out of the laundry room with a basket of folded
clothes. She walks through the house...

...dropping dish towels in THE KITCHEN... hand towels in THE
BATHROOM... and finally turning into...

125   **INT. RICK'S BEDROOM, ARMSTRONG HOUSE - LATE AFTERNOON**   125

Janet puts the basket on the bed, moves to put a few shirts in
his dresser drawer... and SOMETHING CATCHES HER EYE.

REVERSE TO the Whites' house, Pat's car in the driveway. Pat
is staring down into her open trunk. She doesn't move.

Janet, DISCONCERTED, wipes off her hands and heads outside.

126   **EXT. ARMSTRONG HOUSE/WHITE'S HOUSE - MOMENTS LATER**   126

Janet, a jacket on, walks across the street, approaches Pat.
We note fallen autumn leaves in a yard that needs raking.

JOSH: This scene is a fictionalization, but it reflects the attitudes Neil and the MSC heads had about the LLTV.

JIM: MSC Director Bob Gilruth and MSC's Director of Flight Operations Chris Kraft both felt that it was only a matter of time before a fatal LLTV crash. They were ready to eliminate the training, but the astronauts, especially Neil, argued that the LLTV gave the astronauts by far the best simulation of flight in the lunar environment.

JOSH: To emphasize how close this was to a fatal crash, we took some liberty with Neil's injuries—we gave Ryan some bad cuts/bruises on his face. In reality, other than badly biting his tongue upon ejection and getting a case of chiggers, Neil had no obvious injuries. Of course, the 14 g's of the Weber ejection seat would have caused soreness for days. And, again, we wanted the audience to viscerally feel how close this was to a fatal accident.

THIS PAGE: To simuate Neil's ejection from the LLTV Ryan Gosling and his stunt double (Adam Hart) were both ejected into the air and dragged across the ground with the parachute.

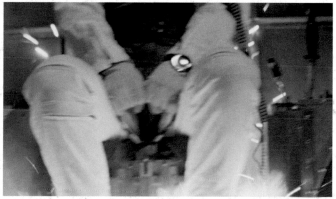

JIM: Although this scene is not based on any specific anecdote, it does underscore how tragic Ed's loss was and the toll this must have taken on his wife Pat and their two children, Bonnie and Ed Jr.

JOSH: We talked about this with Rick and Mark, as well as Ed and Pat's daughter Bonnie. Olivia Hamilton, who played Pat, also spent some time with Bonnie. From all sources, it was clear that Pat was devastated in the wake of Ed's death. Rick remembers being over at the White's house after and said that Pat was "checked out." We wanted to show this (and what must have been its profound impact on Janet) in a more visceral, haunting way than has been done in the past.

First Man   POST CONFORMED BLUE   80.

                    JANET
          Pat? Pat. You okay?

At last, Pat looks up. Eyes VACANT.

                    PAT
          Yes.

The trunk is **EMPTY**. Janet sees Eddie in the window, then
glances at Pat, now staring off into the distance. Janet
SWALLOWS her emotions.

                    JANET
          Why don't we go inside.

She closes the trunk. And gently leads Pat inside.

127   **INT. KITCHEN, ARMSTRONG HOUSE - LATER**   127

Janet walks in. **DEEPLY UNSETTLED**, *more **FRAGILE** than we've
seen*, she pulls a pack of cigarettes from the drawer...

She tries to light up, but her hands are **SHAKING**.

At last she manages, takes a shaky drag, trying to settle
herself. A beat, then she hears the door. She looks up...

Neil walks in, hurries down the hall, **WIPING** past the kitchen.

HOLD ON Janet, **CLOCKING** the bruise on his face.

She starts to follow... then stops to put the cigarette out.
As she rushes after Neil, we **FOLLOW HER** down the hall into --

128   **INT. BEDROOM, ARMSTRONG HOUSE - CONTINUOUS**   128

Neil's changing his shirt.

                    JANET
          Are you okay?
           (off his face)
          Jesus.

                    NEIL
          I'm fine.

                    JANET
          Look at your face.

Janet's concerned, but Neil **AVOIDS** her gaze, heading past her,
back into --

First Man   POST CONFORMED BLUE   81.

A129 **INT. HALL/KITCHEN, ARMSTRONG HOUSE - CONTINUOUS**   A129

Neil moves down **THE HALL** towards the kitchen. He tries to act
normal, tucking in his shirt...

...but THE BOYS run in, excited Neil's home early.

                    RICK
          Dad, wanna come play?

Neil, flustered, doesn't answer. Janet pulls the boys back as
he pours himself a glass of iced tea.

                    JANET
          Ricky, boys, go back, go back and,
          go back and play your game.

          RICK   JANET
His face...   I know, I know. Dad's fine.
                              Go back, go back and play...

They back out... but Janet turns back to Neil and can't stop
herself. She walks towards him.

                    JANET
          What happened?

Neil hesitates, it's suddenly **STIFLING**. He can't be there.

          NEIL   JANET
I, uh...   (reaching for his face)
                              Jesus.

                    NEIL
          I just remembered, I left something
          at the office.

He walks past her, grabbing his briefcase. She's stunned.

                    JANET
          Do you know what time you'll be
          back?

But he walks out the door. **HOLD ON** Janet for a moment,
DESPERATELY STRUGGLING to find her balance...

B129 **EXT. ARMSTRONG HOUSE - CONTINUOUS**   B129

Neil gets into the car and backs down the drive way. As he
pulls off, leaving the house behind, we hear **DRUM BEATS**...

129   **OMITTED**   129

JIM: After the LLTV accident (and being checked out by a medic), Neil went back to his desk like nothing had happened. Al Bean returned from lunch and saw Neil at work in the office the two men shared. A little later, Bean overheard a group of colleagues talking about Neil's accident and told them it couldn't be true—Neil was at his desk. But, of course, that was just Neil.

JOSH: Damien and I love that story, but we'd already demonstrated how unflappable Neil was. We wanted a more human moment. And Neil didn't stay at work all day; Rick told us when he got home at 3.30 p.m., his father was there. When Rick asked why Neil was home early, Janet told him, "Don't talk to your father, he bit his tongue real bad."

The famous Al Bean story also undercuts one of the larger thematic points Damien and I are trying to make—that no matter how publicly calm and stoic these men were in the face of danger and tragedy, these men were human. And there had to have been an emotional cost.

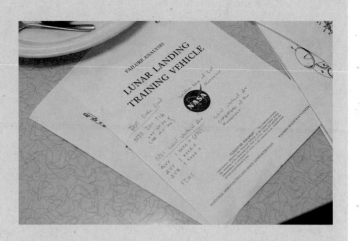

We obviously embellish a bit here. We portray Neil more obviously brusque and out of sorts than he might have been and Janet more obviously upset. But we wanted to get across how bad this accident could have been, how terrifying that must have been for Janet in the wake of the deaths that had come in the past year. And, on some level (even if deeply, deeply buried), how unsettling this must have been for Neil as well.

"SURE, THERE ARE DANGERS AND OUR BUSINESS IS TRYING TO FIND OUT WHERE THE DANGER SPOTS ARE AND MAKE THEM LESS DANGEROUS. WE SPEND ALL OUR TIME DOING THAT. IT'D BE SILLY TO SAY THAT WE DON'T THINK OF THE DANGERS BECAUSE THAT'S WHAT WE DO ALL THE TIME."

NEIL ARMSTRONG

First Man   POST CONFORMED BLUE   82.

A130 **NASA PROTEST MONTAGE**   A130

In a series of <u>QUICK CUTS</u>, a number of public figures and ordinary Americans **CRITICIZE** NASA and the TAX DOLLARS spent on the attempt to send a man to the Moon. This takes us to --

130 **EXT. PROTEST, CAPE KENNEDY - DAY**   130

Over a hundred **PROTESTORS**, many camped out, some playing music. A couple news vans. We **PAN OVER** a SPATE OF SIGNS...

"<u>How many must we sacrifice?</u>" "GRISSOM, WHITE, CHAFFEE, SEE, BASSETT, FREEMAN, WILLIAMS, GIVENS, LAWRENCE."

"<u>Billions for space, but pennies for the hungry?</u>"

Beside this last sign, we find a drummer and an African American man, GIL SCOTT HERON, singing into a microphone.

> GIL SCOTT HERON
> *A rat done bit my sister Nell*
> *with Whitey on the Moon.*
> *Her face and arms begin to swell*
> *and Whitey's on the Moon.*

Heron continues singing as we **INTERCUT WITH** --

INT. F-1 MANUFACTURING PLANT, ROCKETDYNE - CANOGA PARK, CA - DAY

Techs oversee **MOLTEN STEEL** poured into a mold. We don't see what they're building at first...

> *GIL SCOTT HERON (V.O.)
> *I can't pay no doctor bill,*
> *but Whitey's on the Moon.*
> *Ten years from now I'll be paying still,*
> *while Whitey's on the Moon.*

ANGLE ON a familiar **F-1 ROCKET ENGINE** from the Saturn V rocket as techs spray it with water, cooling it down.

CAPE KENNEDY PROTEST

The mixed crowd gathers in, listening to Heron as he sings in front of a purple peace sign.

> GIL SCOTT HERON
> *You know the man just upped my rent last night.*
> *Cause Whitey's on the Moon.*
> *No hot water, no toilets, no lights.*
> *But Whitey's on the Moon.*

We now see the familiar <u>VAB BUILDING</u> towering in the distance. The music takes us to...

JIM: 1968 was one of the most volatile years in American history, with the assassinations of Martin Luther King Jr. and Robert F. Kennedy, unrest on college campuses, widespread protests against the Vietnam War, and much more. So, it is no surprise that there would be protestors at the launch of Apollo 8 (which was followed by a more organized protest at the launch of Apollo 11, led by Reverend Ralph Abernathy).

JOSH: As I mentioned earlier, we wanted to get across the changing times, the sense that the world was shifting around these astronauts, along with the political appetite for their mission.

JIM: The use of African-American poet/musician Gil Scott-Heron's 'Whitey on the Moon' though not written or performed until 1970, gives a voice to the sincere concerns that many Americans had about giving priority to the Moon mission when millions of poor Americans suffered from hunger, homelessness, dispossession, and poor education.

ABOVE: Protesters gather and rail against the cost of the lunar mission in money and lives.
BELOW: Gil Scott Heron (Leon Bridges) sings 'Whitey on the Moon.'

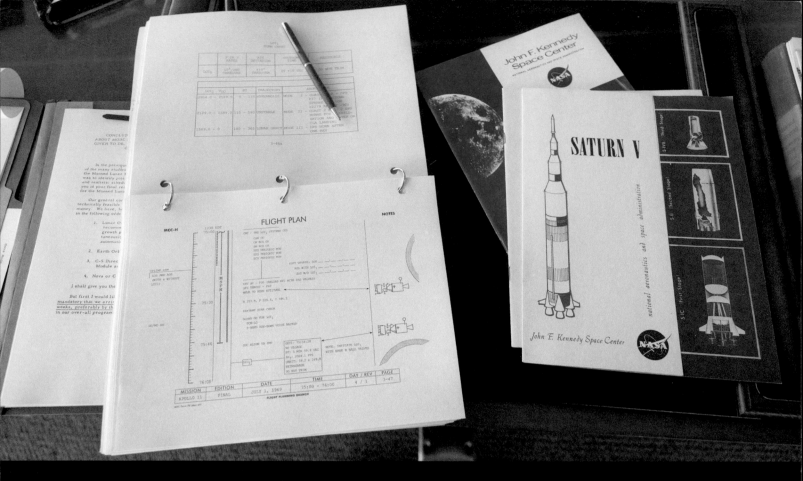

JOSH: Neil and Buzz were backup on Apollo 8, this felt like a good time to reintroduce them, along with Mike Collins, who was a CapCom on the mission (Collins had originally been chosen for the prime crew, but after being diagnosed with a herniated disc that required surgery, was replaced by Jim Lovell).

JIM: Collins dubbed the three of them as "amiable strangers." But we do know of a few less than amiable moments between Neil and Buzz. There was an argument over a LM simulator crash at KSC, not to mention some friction over the issue of who would be first man on the Moon ("first out"), which I go into in some detail in my book.

JOSH: Yes. And, as we pointed out earlier, given the personality differences between these two men, it's hard not to imagine any number of smaller moments like this one, even for someone as non-confrontational as Neil.

JIM: That said, I wonder about Buzz's language here. What was untested was not the Saturn V rocket but a manned launch of the entire stack of rocket and spacecraft. So I'm not sure Buzz would have called it an "untested rocket."

JOSH: We're pushing this a bit. But this launch was rushed for the political reasons Buzz mentions. And, as Jeffrey Kluger writes in his great new book, *Apollo 8 · The Thrilling Story of the First Mission to the Moon*, there were any number of untested/risky facets of the mission. The Saturn V was notoriously buggy; the Apollo 8 crew were the first humans to pass through the Van Allen belt; and the first to attempt trans-lunar and then lunar orbit injection (missing on either would have sent them hurtling off towards the Sun). It was an incredibly bold and risky mission. We're amping Buzz's language to try to get that across, and to reinforce the kind of unvarnished truth-teller Buzz could be in these moments.

JIM: It is true that guys were starting to look at the rotation and wonder who was going to be first on the Moon. And the LLTV training would have been some indication of that.

JOSH: One other note—the reflection of Apollo 8 in the window of launch control is a bit of a cheat; it seems as if the launch pad is closer than it actually is. Due to production/ budget constraints, doing the launch like this was the only way to keep the scene in the movie. And to get across how impressive the launch was.

First Man   POST CONFORMED BLUE   83.

131 **EXT. VEHICLE ASSEMBLY BUILDING, KENNEDY SPACE CENTER - SAME TIME**

CLOSE ON the a huge **SATURN ROCKET**... rolling out of the VAB.

> GIL SCOTT HERON (V.O.)
> *I wonder why he's upping me?*
> *Cause Whitey's on the Moon?*
> *I was already giving him like fifty a week.*
> *With Whitey on the Moon.*

The drums continue as we **REVEAL** a massive <u>APOLLO CRAWLER-TRANSPORTER</u> slowly moving the SATURN towards the launch pad.

> MIKE COLLINS (O.C.)
> Jesus, that's a big mother.

MIKE COLLINS, 38, is with Neil, Buzz, Lovell and others, just beyond the Crawler-Transporter, staring up at the rocket.

> BUZZ
> It'll go up like a half kiloton a-bomb if it blows.

The guys look at Buzz; will he take a hint? He doesn't.

> *BUZZ
> It's a political rush job. Congress wouldn't fund us to come in second. Why else would NASA risk sending a new rocket to the moon on its first manned launch?

Lovell stares at him. Then --

> LOVELL
> Thanks for the insight, Buzz.
> Always a pleasure with you.

Lovell walks off. Mike shoots a look at Buzz.

> BUZZ
> Doesn't matter. He's not in the
> lunar lottery.

> MIKE COLLINS
> And you are?

> BUZZ
> The only guys they let on the LLTV
> since Neil's accident are the ones
> who might land.
> (MORE)

> BUZZ (CONT'D)
> That's Neil or Conrad and I'm backup
> with Neil, so...

> COLLINS
> So you think you're going to the
> Moon.

> BUZZ
> It's been up for grabs since Gus
> died.
> (then)
> I'm just saying what you're
> thinking.

> NEIL
> Well, maybe you shouldn't.

Mike turns towards Neil, surprised. We **PUSH IN** on the HUGE
ROCKET and **PRELAP** --

> LAUNCH CONTROL (**PRELAP**, OVER PA)
> *All systems go on Apollo 8...*

132   <u>**INT. PRIVATE VIEWING ROOM, KSC - DAY**</u>   132

TIGHT ON Neil, Buzz, FRED HAISE and a few military uniforms
THROUGH A GLASS PANE. Neil stares out...

> **Apollo 8 Launch**
> **December 21, 1968**

...in the glass, the reflection of a FAMILIAR ROCKET.

> LAUNCH CONTROL (OVER PA)
> *...man's first attempt to orbit the*
> *Moon. The engines are armed...*

SLOWLY PUSH IN ON NEIL, as flames from the launch begin to
light up his face. The **RUMBLE** of the rocket **GROWS**, but our
focus remains on Neil...

> LAUNCH CONTROL (OVER PA)
> *4, 3, 2, 1, 0, we have commit, we...*

...until the <u>rocket's **ROAR**</u> drowns out the PA. It's like
nothing we've ever heard. We hear nothing else as we **PUSH IN**
on Neil, his eyes moving up, following the rocket until we...

RACK FOCUS to THE GLASS, the reflection of Apollo 8 racing
towards the heavens. **RACK BACK** to Neil, **CLOSE ON** his eyes.

He walks off and we... **REVEAL** Deke watching Neil walk away.

"IT'S REALLY ABOUT BUILDING THE EMOTIONAL JOURNEY SO THAT THE SUCCESS OF LANDING ON THE MOON BECOMES AN EMOTIONAL RELEASE FOR NOT ONLY THE CHARACTERS BUT THE AUDIENCE."

WYCK GODFREY, PRODUCER

First Man   POST CONFORMED BLUE   85.

133   **INT. BATHROOM, KSC - CONTINUOUS**   133

Neil washes his hands. Methodical.

                    DEKE
              Helluva launch.

Neil looks up, sees Deke's walked in behind him.

                    NEIL
              Yeah.

                    DEKE
              Everything stays on track, Eleven's
              gonna be the landing. I talked to
              Bob, everyone's in agreement, we'd
              like you to command.

Neil's pleased, but doesn't acknowledge how big this is.

                    DEKE
              The doc cleared Collins, he's the
              best Command Module Pilot available.

                    NEIL
              And Buzz for LM Pilot.

                    DEKE
              That's the rotation. But no one
              would fault you if you'd rather take
              Lovell.

Huh. Helluva choice. A beat.

                    NEIL
              Let me think about it.

Off Neil, considering we **PRELAP** --

                    CRONKITE (**PRELAP**, ON TV)
              It looks like a red dot with...

INT. LIVING ROOM, ARMSTRONG HOUSE - MAY 26, 1969 - 11:52 AM

CLOSE ON a TV. Footage of the Apollo 10 re-entry.

                    CRONKITE (ON TV)
              *...a long tail, a long plume, that's
              got to be the spacecraft, that has
              to be Apollo 10 re-entering.*

FIND Neil, on a couch, watching. **PUSH IN** on him. PROCESSING.
Not celebratory. Just on to what's next.

JOSH: One thing that we try to subtly point to here is that, at this time, it wasn't certain that Apollo 11 would land on the Moon.

JIM: That's true. If anything had gone wrong with 9 or 10, the landing could have been pushed to 12. Pete Conrad might have been the first man on the Moon.

JOSH: Imagine that!

JIM: Neil was always very humble about his role, noting that he was just the point of a very long spear made up of thousands of people and hundreds of thousands of hours.

JOSH: So, the real moment Neil knows Apollo 11 is definitively going to attempt to land on the Moon is when Apollo 10 returns safely.

JIM: You make an interesting choice in how you play that scene.

JOSH: Obviously, it's an exciting moment. But it also means a challenging journey ahead, I might even say a harrowing one for Janet. After all, Neil's only other spaceflight had been excruciating for her—and that one was a lot simpler and a lot shorter.

JIM: By the way, the bit about Buzz is based on an actual conversation that neither Neil nor Deke disclosed until Neil discussed it with me during one of our interviews for my book. The revelation that Deke gave Neil the option of taking Lovell instead of Buzz was stunning. As scripted here, Neil told me he asked Deke if he could think about it, then came back to Mission Control the very next day and told Deke he thought Lovell deserved to have his own command.

JOSH: At the moment, this interchange isn't in the cut. I've kept it here because I find it fascinating as well, but it's a bit off our main storyline so I doubt it will be in the final film. I've found editorial can be a pretty unforgiving place—diversions from narrative you can get away with on the page often don't work nearly as well on screen.

First Man   POST CONFORMED BLUE   86.

> CRONKITE (ON TV)
> *And so, the flight of Apollo 10 has performed the major function of its mission. It has proved through these daring three astronauts that all of the systems work properly and that there should be no reason why man cannot, perhaps as early as July, land on that picked spot on the Moon's equator...*

Now we **FIND** Janet, watching from the hall, far from Neil.

> CRONKITE (ON TV)
> *These are sailors of the sky and what we've seen and heard today make the great ocean voyages of the earthbound seem, well, earthbound indeed.*

The moment **LANDS** on each of them separately. At a loss, she walks back down the hall and into --

A135   **INT. BEDROOM, ARMSTRONG HOUSE - CONTINUOUS**   A135

She stands there for a moment. Mark peeks in, goes inside.

> MARK
> Mom? What's wrong?

> JANET
> Hmmm?

> MARK
> What's wrong?

> JANET
> Nothing, honey. Your dad's going to the Moon.

Mark takes this in. Not quite processing.

> MARK (O.C.)
> Okay. Can I go outside?

> JANET
> Sure.

He runs off. We **STAY WITH** Janet, then we **SMASH TO** --

135   **INT. MOVIE THEATER, MANNED SPACE CENTER - EARLY MORNING**   135

THREE MEN in **GAS MASKS** are led into a theater packed with reporters, pads and cameras ready. It's ODD.

Apollo 11 Pre-Flight Press Conference
July 5, 1969

First Man   POST CONFORMED BLUE   87.

The men are led on stage to a **TALL, THREE-SIDED PLASTIC BOX**.
They sit, take off the masks. It's Neil, Mike Collins... and
<u>BUZZ</u>. Deke, off to one side, nods to Neil.

> NEIL
> We're here today to talk a bit about
> the forthcoming flight. But we're
> able to talk about it because of
> previous flights. Every flight had
> new objectives and left us with very
> few additions to be completed.
> We're very grateful to those people
> who made it possible for us to be
> here today.

Neil sits, <u>no trace of emotion</u>.

> DEKE
> Alright, we'll take some questions.
> Jim?

> REPORTER #1
> Neil, when you learned you were
> going to command this flight, were
> you surprised? Overjoyed?

> NEIL
> ...I was pleased.

> REPORTER #1
> Okay, but how would you compare this
> feeling to winning an automobile? Or
> being selected as an astronaut?

> NEIL
> (pause)
> I was pleased.

The camera **PUSHES IN** on Neil as the questions come **FASTER**...

> NEWSWEEK REPORTER
> Neil, were you aware that Ralph
> Abernathy is planning a protest for
> the day of the launch?

> NEIL
> No. I wasn't.

More questions come, we're **TIGHTER** on Neil now.

> REPORTER #2
> Neil, if it does turn out, you'll go
> down in history.
> (MORE)

JOSH: We open with an abbreviated version of the actual statement Neil gave at the beginning of this press conference on July 5, 1969.

JIM: It's a good place to start, as this idea of standing on the shoulders of the work, the men and women that came before is a recurring theme in Neil's comments about the Moon landing. Now, the press conference was actually moderated by PAO Brian Duff.

JOSH: Yep, we take a little creative license here—the audience is familiar with Deke and he seems to be a natural facilitator. Plus, we all love Kyle Chandler and he helps warm up what is a somewhat distancing scene and sequence.

JIM: Neil did answer the vast majority of the questions at this press conference—27 of the 37 asked. And on a few occasions, Buzz, unsolicited, added to Neil's comments, as he does here on this page of the script. But not all these questions came from this press conference.

JOSH: No, we took several of them from other interviews Neil gave in the lead up to Apollo 11. In fact, the only question we made up is the one about Ralph Abernathy.

JIM: I know you've cut this line from the film, but Abernathy did indeed march four mules and 150 members of the Poor People's Campaign for Hunger as close as they could get to KSC.

JOSH: We also embellished the Buzz lines a bit, but Buzz did tell reporters he was bringing some of his wife's jewelry to the Moon in his PPK. In fact, all the men brought gold olive branch pins to give to their wives upon returning.

Finally, I think it's worth mentioning that the more fuel line is verbatim.

JIM: Yes, and quintessential Neil.

LEFT: The Apollo 11 crew before the pre-launch interview. The masks aren't featured in the film, but the real crew was indeed made to wear them to avoid contracting anything that might scratch them from the flight. The box around the men (with fans blowing out towards the press) was set up for a similar reason.

First Man   POST CONFORMED BLUE   88.

                REPORTER #2 (CONT'D)
What kind of thoughts do you have
about that, when a thought hits you -
- 'Gosh, suppose that flight is
successful' --

                NEIL
We're planning on that flight being
successful.

                REPORTER #2
Uh, I just meant, how you feel about
yourself being a part of history?

Neil hesitates... and in that moment <u>Buzz jumps in</u>.

                BUZZ
I think I can shed some light here.
It's a responsibility, but it's
exciting to be the first. Even my
wife is excited. She keeps slipping
jewelry into my PPK.

Some laughter. Neil looks **IRRITATED.**

                REPORTER #2
You're planning to take some of her
jewelry to the Moon, Buzz?

                BUZZ
Sure. What fella wouldn't want to
give his wife bragging rights?

Laughter. Buzz smiles, enjoying the limelight.

                REPORTER #3
Neil, will you take anything?

               *NEIL
If I had a choice, I'd take more
fuel.

A few chuckles, but not many.

                DEKE
Alright, next question.

Off Neil, **UNCOMFORTABLE, CUT TO --**

136 **<u>OMITTED</u>**   136

137 **<u>OMITTED</u>**   137

THIS PAGE: Neil (Ryan), Buzz (Corey), and Mike Collins (Lukas Haas) at the Apollo 11 pre-flight press conference.

"THERE IS A VERY REAL DANGER THAT OUR OWN GENERATION MAY TOO SOON FORGET THAT PROGRESS HAS A PRICE WHICH SOCIETY MUST BE WILLING TO PAY."

NEIL ARMSTRONG

JOSH: Okay, I'm gonna jump out and say it. There's some dramatic license taken here. But, this scene is grounded in fact. We talked to Rick and Mark, and while Neil was incredibly even tempered and non-confrontational, Janet could yell (and swear) on occasion. Janet also spoke with you, Jim, and others about the challenge she faced trying to get Neil to engage emotionally with her and the boys. "I just couldn't live with the personality anymore," was how Janet put it, when explaining why she eventually left Neil.

138   <u>**INT. BEDROOM, ARMSTRONG HOUSE - NIGHT**</u>   138

8pm. Janet walks in to find Neil. <u>PACKING</u>, methodical.

She stands there for a moment. Watching him. GIRDING herself
for a confrontation. A deep breath, then quietly --

                    JANET
          I thought you were gonna talk to the
          boys.

                    NEIL
          Well, what did you want me to say?

                    JANET
          What do you want to say? You're the
          one that's going away.

He doesn't say anything. Just keeps packing.

                    JANET
          You're just killing time until you
          can get in the car.

Neil pauses, then walks past her. **CUT TO** --

A139   <u>**INT. NEIL'S OFFICE, ARMSTRONG HOUSE - MOMENTS LATER**</u>   A139

Neil packs up his briefcase. Janet walks in.

                    *JANET
          Neil, I need you to talk to the
          boys.
               (then)
          Can you hear me? I need you to talk
          the boys. What are you doing?

                    NEIL
          I'm going to work.

And now, her anger **CRESCENDOES**, overtaking her...

                    JANET
          Well, just stop it. Just stop, just
          stop <u>packing</u>.

Janet grabs his briefcase, hurls it onto the floor. Neil
looks up at her. She slams the office door shut.

                    JANET
          What are the chances you're not
          coming back? What are the chances
          this is the last time the boys are
          gonna see you?

First Man   POST CONFORMED BLUE   90.

> NEIL
> I can't give you an exact number.

> JANET
> I don't want a fucking number, Neil!
> It's not zero. Is it? Is it?

Neil hesitates. Looks down...

> NEIL
> No.

> JANET
> No, it's not. Pat doesn't have a
> husband, those kids, they don't have
> a father. What are the chances
> that's gonna be Ricky and Mark?
>   (then)
> You're gonna sit them down now, both
> of them, and you're gonna prepare
> them for the fact that you might not
> ever come home. You're doing that.
> You. Not me. I'm done.
>   (pause, then)
> So you better start thinking about
> what you're gonna say.

Janet turns and walks out. Off Neil, **CUT TO --**

139   **INT. DINING ROOM, ARMSTRONG HOUSE - NIGHT**   139

Neil and Janet sit with Rick and Mark in pajamas. Rick is
slightly disengaged. A long silent beat. Then --

> MARK
> Jimmy asked what you're going to say
> when you get on to the Moon.

> NEIL
> Well, we're not sure we're gonna get
> on to the Moon, a lot of things have
> to go right before that happens.

Another awkward silence. Then --

> MARK
> How long will you be gone?

> NEIL
> Well, we launch in ten days. We'll
> be up for eight. And then about a
> month in quarantine.

JIM: I do feel the scene is honest in portraying the intensity of Janet's feelings when it came to Neil not engaging directly enough with his family. And in one of my interviews with Janet for my book, Janet recalled talking to Neil just before he left for the Apollo flight, asking him to talk to the boys. There was a possibility that he might not survive, and Janet wanted Neil to address that with Rick and Mark. But she didn't mention any kind of blow out.

JOSH: Neil and Janet had been through so much at this point. We wanted to try to show the cost of all the loss and all the stress on the family in a visceral way. As you say in your book, Jim, the wonder is not that they eventually divorced, but that their marriage lasted as long as it did.

> **"THE CAMERAMAN IS FOLLOWING YOU AND WITH YOU, HE'S PART OF THE PERFORMANCE AS MUCH AS YOU ARE."**
> CLAIRE FOY, 'JANET ARMSTRONG'

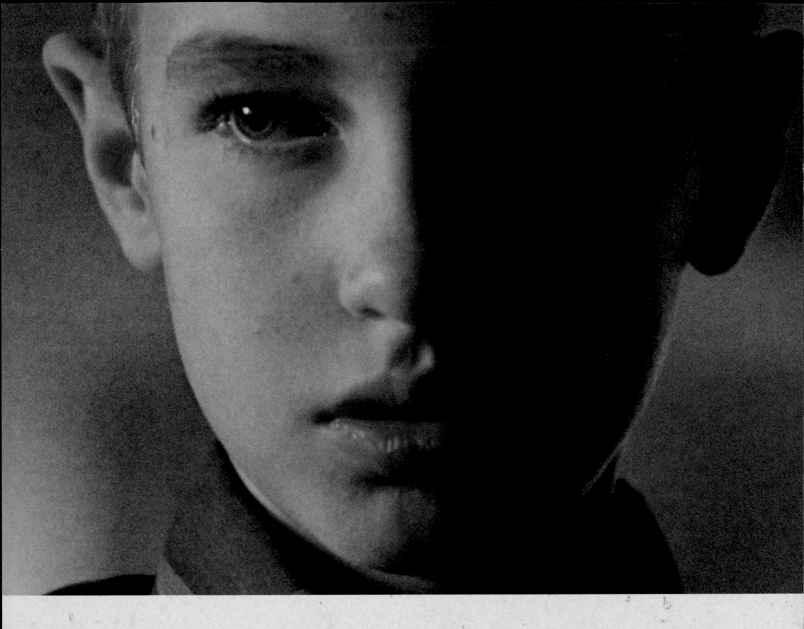

JIM: Janet didn't recall much about this conversation. Specifically, in terms of Neil discussing the possibility he might not return, she said: "I don't think that went very far. I don't know what he might have said to them."

JOSH: Rick and Mark proved helpful here. Rick told us the conversation was in the dining room. Rick also told us that he wasn't that worried about Neil coming back—he believed that if anything went wrong then they'd figure it out. But when we told Rick that we thought it was an important question he gave us the language that Neil used at the outset of the conversation and when he started the conversation about the flight: "We have confidence in the mission. There is some risk in it. We have every intention of coming back."

The handshake was something we played with, trying to get at the oddness of the moment, the challenge of growing up as the son of Neil Armstrong. Sure, it's amazing to have a dad who's the first man on the Moon, but it's also gotta be hard in many ways.

First Man   POST CONFORMED BLUE   91.

> MARK
> What's quarantine?

> NEIL
> We'll be in isolation. To protect
> in case we, uh, carry any diseases
> from the lunar surface, or something
> of that nature. It's not likely,
> but it's a precaution.

> MARK
> So you won't be here for my swim
> meet?

> NEIL
> No. I'm sorry.
>   (beat)
> Does anyone have any more questions?

> RICK
> Do you think you're coming back?

Neil looks up at him.

> *NEIL
> We have real confidence in the
> mission. And there are some risks,
> but we have every intention of
> coming back.

Not comforting. Rick's old enough to read between the lines.

> RICK
> But you might not.

> NEIL
> ...That's right.

Neil shifts, uncomfortable.

> JANET
> Okay. Okay, time for bed.

They get up. Mark gives Neil a hug, then scurries off. Rick
walks over, then holds out a hand.

Neil hesitates, then shakes hands with Rick.

It's heartbreaking in its distance, its formality. Off Janet--

140   **INT. GILRUTH'S OFFICE, MANNED SPACE CENTER - NIGHT**   140

Gilruth sits, reviewing some papers. Deke walks in.

First Man   POST CONFORMED BLUE   92.

                              GILRUTH
                 White House sent down a contingency
                 statement. You mind if I...

Deke nods. Gilruth picks up the speech. **"CONTINGENCY
STATEMENT IN EVENT OF MOON DISASTER."** And starts to read...

                              GILRUTH
                 *Fate has ordained that the men who
                 went to the Moon to explore in peace
                 will stay on the Moon to rest in
                 peace. These brave men, Neil
                 Armstrong and Edwin Aldrin, know
                 there is no hope for their recovery.*

141 **INT./EXT. ARMSTRONG HOUSE – NIGHT**   141

WIDE on the house. A GOVERNMENT SEDAN idles. Neil comes to
the door, duffel and a briefcase. Janet's behind him.

                              GILRUTH (V.O.)
                 *They will be mourned by their
                 families; they will be mourned by a
                 Mother Earth that dared send two of
                 her sons into the unknown...*

Neil gives her a peck on the cheek, walks to the sedan, hands
off his duffel and, without looking back, gets in...

                              GILRUTH (V.O.)
                 *Others will follow, and surely find
                 their way home. But these men were
                 the first, and they will remain the
                 foremost in our hearts.*

HOLD ON Janet. **ALONE.**

142 **INT. GOVERNMENT SEDAN, ARMSTRONG HOUSE – SAME TIME**   142

Neil sits back. He pulls out a mission briefing, starts to
read as the driver gets back in the car.

                              GILRUTH (V.O.)
                 *For every human being who looks up
                 at the Moon in nights to come will
                 know there is some corner of another
                 world that is forever mankind.*

The sedan pulls off and <u>tension LEAVES Neil's face.</u> He looks
out at the houses slipping away... and we see **RELIEF.**

<u>The mission has begun.</u>

outside of the White House. In fact, the memo was only widely circulated after it surfaced in the National Archives around the 30th anniversary of the Moon landing. So, while the memo was real, neither Gilruth nor Deke would have been privy to it.

JOSH: One of the key challenges in telling this story is getting an audience to forget what they know. Everyone knows Neil Armstrong landed on the Moon and returned home safely. But we need to take people back to the moment when that was far from a given, when there was a substantial enough chance of disaster that a busy White House speechwriter would be given this assignment. That's why Damien and I took some creative license here. We wanted to hear these words as Neil leaves home.

"YOU GET SO INTENTLY INVOLVED, AND THERE IS ALWAYS A STRAIN... DEFINITELY A STRAIN ON THE WIFE AND FAMILY."

JANET ARMSTRONG

First Man   POST CONFORMED BLUE   93.

143   **INT. GILRUTH'S OFFICE - NIGHT**   143

                         GILRUTH
               Prior to the statement, the
               President will telephone each of the
               widows-to-be. A clergyman will
               adopt the same procedure as a burial
               at sea, commending their souls to
               'the 'deepest of the deep.
                    (then, looking up at Deke)
               Any thoughts?

A beat. The speech has had an impact. Deke pushes it aside.

                         DEKE
               Sounds fine.

144   **EXT. LAUNCHPAD, KENNEDY SPACE CENTER - 4:15 AM**   144

HUGE FLOODLIGHTS **ILLUMINATE** the **SATURN V** STEAMING on the pad.

                4:15 AM, Kennedy Space Center
                        July 16, 1969

The **TERRIFYING GIANT** casts shadows over TECHS hooking up **HOSES**
for fuel, oxygen and nitrogen. **MUFFLED THUMPING** takes us to --

145   **EXT. KENNEDY SPACE CENTER - 4:15 AM**   145

Wide on a building. DARK, save for a a few lit windows.

146   **OMITTED**   146

147   **INT. MESS HALL, KENNEDY SPACE CENTER - 5AM**   147

Neil works through steak and eggs, incessantly looking over a
map of the lunar surface. Mike and Buzz sign **POSTMARKED**
ENVELOPES with **COMMEMORATIVE STAMPS**. A sketch artist sits
beside Deke, making quick work of all three of them. It's
still dark outside.

A beat, then Deke catches Neil's eye. It's time. Neil gives
a half nod, stands. The others follow suit and we **CUT TO** --

148   **EXT. LAUNCHPAD, KENNEDY SPACE CENTER - 5:30 AM**   148

Personnel scurry around the **HUGE ROCKET**, doing final checks.

            5:30 AM (*T minus 4 hours, 3 minutes*)

A **BUZZER**, then a CALL for non-essential personnel to leave the
pad. We spot **HUGE FUEL TANKS** beside the pad... and **MASSIVE**
HOSES filling the rocket with **LIQUID OXYGEN** and **HYDROGEN**.

JOSH: So we tried to get as close as possible to recreating the breakfast and the events that lead up to launch. NASA had sketch artist Paul Calle on hand to create portraits of the astronauts so we have his son, artist Chris Calle making sketches of our astronauts. To make sure we suited up the guys right, we had suit techs Al Rochford and Ron Woods with us—Ron had suited Buzz up on Apollo 11. And we shot our Apollo 11 astronauts walking out of the Manned Spacecraft Operations Building (now the O&C or Operations and Checkout building) at KSC in the same place the actual Apollo 11 astronauts did and had them get into the actual van that drove Neil, Mike, and Buzz to the Launch Pad.

JIM: One little nit pick—the commemorative postal covers the guys signed and took to the Moon had already been packed aboard the spacecraft. There were other "insurance covers"—astronauts could not obtain much life insurance, so they signed these, on the presumption they would become highly valuable in the event of their deaths. But they were signed weeks before. One of our consultants, Robert Pearlman, is an expert on space memorabilia, and he confirmed that for us.

JOSH: Yeah, we loved the detail and this was the best place to incorporate it.

As we moved into the spacecraft, Al Worden, Command Module Pilot on Apollo 15, was on hand and very helpful. We also spoke to both Mike and Buzz and they were kind enough to share what they remembered.

We wanted a direct visual to show Neil's focus on the Moon. So did Neil—according to Rick and Mark when people would ask him if it was hard to get to the Moon, he'd say "No, I could see it the whole way."

THIS PAGE: The *First Man* team recreates this breakfast scene.

As the techs check the hoses, **INTERCUT WITH** --

149   <u>**INT. THE "SUITING UP" ROOM, KSC – SAME TIME**</u>   149

Tape **SEALS** up a <u>URINE COLLECTION DEVICE</u> around a man's waist.

Hands pull on a basic set of **COTTON LONG UNDERWEAR**.

A THERMAL SKIN is **VELCROED** on top of that.

METALLIC RINGS **SNAP ON** one glove, then another.

A COMMS (SNOOPY) CAP chin-strap is **SNAPPED** in place.

The WIRE from the cap is **PLUGGED INTO** the top of the suit.

An AIR NOZZLE is **TWISTED ONTO** the BLUE PORTAL on a spacesuit.
We hear the **HISS OF OXYGEN** as...

A HELMET is **PLACED** on the head of a <u>SILENT NEIL</u>.

The helmet **SWIVELS** in place and the sound **FADES**... we hear
only the oxygen's hiss. **PUSH IN** on Neil's eyes, **CUT TO** --

150   <u>**INT. CREW QUARTERS – LATER**</u>   150

Neil, **SUITED UP**, stares straight ahead as he walks down the
corridor... shaking hands with cheering techs. All we hear is
the **HISS** in Neil's helmet as he walks toward DOUBLE DOORS...

The doors swing open and Neil walks through. CUT TO --

A151   <u>**EXT. CREW QUARTERS – CONTINUOUS**</u>   A151

Along with Mike and Buzz, Neil walks to the van, passing the
spectators he (and we) can't hear. He gives a thumbs up,
still staring **STRAIGHT AHEAD**, and we **MATCH CUT TO** --

151   <u>**EXT. GANTRY ELEVATOR, LAUNCH PAD 39A, KSC – LATER**</u>   151

CLOSE ON Neil. The HISS of air pervades.

          6:45 AM (*T minus 2 hours, 48 minutes*)

Neil rides up the side of the **ENORMOUS SATURN V ROCKET** with
Buzz, Mike and Deke. We truly appreciate how **HUGE** the rocket
is. 363 feet tall, 33 feet wide... it dwarfs the world below.

Buzz finds it STAGGERING. Neil is too **FOCUSED** to notice.
The elevator **JERKS** to a halt. Neil picks up his oxygen tank,
leads the others onto --

First Man   POST CONFORMED BLUE   95.

152   **EXT. SWING ARM 9, LAUNCH PAD 39A - CONTINUOUS**   152

Neil walks across the vertiginous **ORANGE STEEL BRIDGE** to the small COMMAND MODULE atop the behemoth. The comm CHATTER is SUBSUMED by the eerie hiss of oxygen and Neil's **INTENSE FOCUS**.

This is a solitary moment. A beat, then Neil walks into --

A153   **INT. WHITE ROOM/APOLLO 11 CSM - CONTINUOUS**   A153

Neil walks into the White Room and a tech walks up. Neil reflexively hands his tank to the tech then pauses...

We **PUSH IN** on Neil's face as the **HISSING** grows louder. A beat, then Neil pushes forward, through the White Room to the waiting spacecraft. He pauses at the open hatch. A beat.

He grasps the overhead handrail and swings himself into the craft, maneuvering into the left hand seat. The techs immediately reach in to hook up his lines and hoses...

The process takes a moment, but Neil just stares at the console. **THE MOON** small in the window.

The lights on the console **FLASH ON**.

We see Buzz to Neil's right and Mike to his far right. The techs close the hatch and we're in...

153   **INT. APOLLO 11 CSM - CONTINUOUS**   153

It's dark, save for a small window. **HOLD ON** the men in the **CLAUSTROPHOBIC** cabin; Neil's window looks into the White Room.

Neil, seemingly oblivious to the stifling atmosphere, starts checks. As the others join, we see **A SERIES OF QUICK CUTS**:

HANDS flipping through the Flight Plan and the Mission Rules.
FINGERS flicking switches, punching buttons.
EYES ticking from books to the console to the clock.

A FINAL FLURRY OF **BUTTONS** AND **SWITCHES** as we arrive at...

                    FIDO (COMMS)
          T minus two minutes and counting.

Nothing to do now but wait. Buzz and Mike twitch with nervous anticipation. Neil keeps his eyes focused. The White Room has been retracted, we now see sky through his window.

PUSH IN on Neil, hand on the ABORT HANDLE. Seemingly **CALM**.

PUSH IN FURTHER on Neil's **EYES**. A WEALTH of **EMOTIONS**. We **HOLD ON THOSE INTENSE EYES** as the call continues...

TOP LEFT: Mike Collins was a great resource on the film. He read the script, gave detailed notes, and shared his knowledge about the Gemini cockpit before Technical Consultant Frank Hughes started the training. Collins came to Kennedy when the cast shot some of the Apollo 11 pre-launch sequence, and it was there that he told the story of how he'd nailed a fish to a wooden plaque and carried it in a bag up to Guenter Wendt in the White Room (a little inside fishing joke between the two of them). Lukas loved the story and he and Damien had the props department find a bag match for Lukas to carry like Mike had.

TOP RIGHT: Model of the Saturn V rocket.

BOTTOM RIGHT: Neil (Ryan Gosling), Buzz (Corey Stoll), and Mike (Lukas Haas) with James R. Hansen.

ABOVE: Neil Armstrong waving in front, and the crew or Apollo 11, head for the van that will take the crew to the rocket for launch to the Moon at Kennedy Space Center.

BELOW: Ryan Gosling, Corey Stoll, and Lukas Haas reenact the famous photo at Kennedy Space Center.

JIM: Many astronauts did not characterize the ride on the Saturn V to be louder or more violent than the ride on the Gemini/Titan. But in his private comments, Neil was very clear that it was both.

As he told me, "In the first stage, the Saturn V noise was enormous, particularly when we were at low altitude because we got the noise from seven and a half million pounds of thrust plus the echo of that noise off the ground that reinforced it. After about thirty seconds, we flew out of that echo noise and the volume went down substantially. But in the first thirty seconds it was very difficult to hear anything over the radio—even inside the helmet with the earphones. It was considerably louder than the Titan. In the first stage, it was also a lot rougher ride than the Titan. It seemed to be vibrating on all three axes simultaneously."

JOSH: And Mike Collins concurred, with typical color, "It was like a nervous novice driving a wild car down a narrow alley and jerking the wheel back and forth."

      FIDO (COMMS)
20 seconds and counting... T LAUNCH DIRECTOR (COMMS)
minus 15 seconds...   *Guidance is now internal...*

       FIDO (COMMS)
   12, 11 10, 9, ignition sequence
   start...

We hear a LOUD **RUMBLING** from far below as the capsule begins
to SHAKE. **HOLD ON** Neil. Steady. <u>Unyielding</u>.

The noise gets **LOUDER**... **DROWNING OUT** the countdown. The
capsule **BUCKS** violently, <u>**vibrating side to side**</u>.

It's **FAR WORSE THAN GEMINI**. Even Neil is <u>STARTLED</u> by the
DEAFENING ROAR and the INTENSITY of the SHAKE, but he **WILLS**
himself to FOCUS, eyes **TICKING** from clock to console...

      FIDO (COMMS)
   Tower cleared.

154 <u>**SMASH OUTSIDE THE CRAFT --**</u>   154

We see BELCHING FIRE, BLACK SMOKE. Ice FALLS, flames RISE --
and the MONSTER LIFTS.

155 **SMASH BACK INSIDE** -- the worst turbulence you can imagine. 155
Everything a blur, severe shakes, the whole craft seemingly
buckling. *Can this be what it's supposed to feel like?*

Neil glances at the clock, monitors the gauges... and we **FEEL**
the rocket **ROLL**, G-forces so INTENSE <u>all three men</u> **FEEL IT**.

More so as blue sky **RAPIDLY TURNS BLACK**... *MUCH FASTER than in*
Gemini or the X-15. Mike reacts, but Neil remains **FOCUSED** on
the instruments as we barely make out COMMS.

      CAPCOM (COMMS)
   Mode 1 Charlie. Go for staging.

BOOM! <u>The men are **SLAMMED** into their seats</u> as we **SMASH TO --**

156-157 <u>OMITTED</u>   156-157

158 <u>**EXT. APOLLO 11, EARTH ORBIT - DAY**</u>   158

From the **POV** of the second stage, we see the CSM pull away.

We **HOLD** for a beat as the CSM gets farther away... then we
gently **FALL**, earth DRIFTING back into view, the sun rising
over it as we **CUT BACK TO --**

"ONE OF DAMIEN'S MISSION
STATEMENTS THAT REALLY
HAD ALL OF US INCREDIBLY
INSPIRED WAS THE
PHYSICALITY OF SPACE FLIGHT.
IT'S PUTTING MEN IN A TINY
CABIN ON TOP OF A MISSILE
AND LAUNCHING THEM INTO
THE UNKNOWN."

ISAAC KLAUSNER, EXECUTIVE PRODUCER

JOSH: We take some license with Translunar Injection. The burn happens behind Earth so the planet doesn't recede in a way that conveys what's happening. We push it visually so the audience will understand we're breaking orbit.

Of course, this kicks off one of the more challenging parts of the script—the trip to the Moon. It's a four-day journey, so there's the obvious question of how to compress this to fit into the third act of a two-hour movie. And the transcript doesn't make it any easier—it's some 600 pages long.

JIM: And hardly the easiest reading in the world.

JOSH: Agreed. Although, unlike Gemini and the X-15, there were more readily available tools to help decipher it. The 'Apollo Flight Journal', for example, was an absolute godsend. But still, trying to figure out what to dramatize after launch was challenging, especially as the trip there was, well, relatively drama free.

JIM: Yes, they had a fairly smooth ride to the Moon. But you found a couple of hiccups.

JOSH: We did. And we focused on them because we wanted to emphasize that, despite the fact that (as Neil liked to say) everything but the landing had already been tested on the prior three Apollo missions, this kind of space travel was still in its infancy and, as such, any journey to the Moon was filled with a number of unknowns.

JIM: I worried about having the script reference the flashes out of window five that Mike noticed just after trans-lunar injection. After all, the audience might think that something was wrong with the spacecraft when it actually was just a visual phenomenon that researchers now believe was caused by cosmic rays interacting with the eyeball and/or the brain.

JOSH: Right. But the crew didn't know what the flashes were at the time. That's why Mike surmises it could be something with the engine. The crew doesn't overreact, but it had to be unnerving. We want the audience to feel the same way, to experience the same disorientation of the crew, to once again

put them back into a place of uncertainty as to the outcome of the mission.

JIM: That's why you have Mike mention the burnt smell when he surveys the probe and drogue after he's docked the CSM with the LM.

JOSH: Yeah, that anecdote comes from Mike's terrific book *Carrying the Flame*. It's one of my favorite astronaut books and he talks about this moment in almost these exact terms there.

JIM: I love that you return to 'Lunar Rhapsody' here.

JOSH: Neil took two pieces of music with him—Dvorak's 'New World Symphony' and *Music Out of the Moon*. I listened to Leonard Bernstein's fantastic recording of the Dvorak constantly while writing (my wife has a habit of listening to a single song while writing a novel, so I learned that from her) so I had originally written in the second movement of the Dvorak piece here. But Damien wanted to call back to 'Lunar Rhapsody', music that was special to Neil and Janet. Maybe because it reminds us how much has been lost or maybe because it shows that even when Neil is further away from Janet than he's ever been, there's still a romantic thread holding these two together. Either way, I didn't need too much convincing.

JIM: There were a couple of Apollo 11 scenes that wound up on the cutting room floor.

JOSH: More than a couple. Even with all my whittling down, I didn't whittle far enough. We had a short Day 2 scene and a short Day 3 scene. More from Day 1 and more from the end of Day 4 and beginning of Day 5. As you can see from the way I've marked up the script, we may yet cut some of the things on these pages. When we watched it all together, we wanted to get to the landing faster. Part of that was a function of length and pacing and part of that was the nature of the material.

JIM: Yeah, you said it earlier that the ride to the Moon was fairly drama free, especially when compared to the actual landing.

159   **INT. APOLLO 11, EARTH ORBIT - DAY**   159

Neil, helmet off, grabs his (floating) helmet and stows it.
Off to the side, we hear Collins and Buzz, helmets off.

>   CAPCOM (COMMS)
>   *Apollo 11, this is Houston.*
>   *Slightly less than 1 minute to*
>   *ignition and everything is GO.*

As Buzz and Neil get to it, Mike glances at the day-lit earth,
maybe thinking that they're about to leave it far behind...

>   NEIL (INTO COMMS)   MIKE COLLINS (INTO COMMS)
> Okay, 59:25 -- and this light   (snapping back to it)
> will go off at 42 --   Time is based on tracking
>   data; let me know when you
>   start it up.

>   NEIL (INTO COMMS)
>   When you feel it, that's when it is.

>   CAPCOM (COMMS)
>   *Apollo 11, you are go for translunar*
>   *injection.*

Mike eyes the **CLOCK** as Neil and Buzz monitor the gauges.

>   NEIL (INTO COMMS)
> Okay, we're operate - 59:59.   BUZZ (INTO COMMS)
>   There we go; thrust.

A **FLASH** out the window and they're **PRESSED BACK** in their seats
again. It's smooth, if JARRING.

>   NEIL (INTO COMMS)
> IGNITION. Call it at 15.   MIKE COLLINS (INTO COMMS)
>   Okay.

Mike watches the earth slowly SLIDE OUT OF THE WINDOW, casting
them into darkness. Buzz turns up the lights on the console
to light their journey into the unknown...

Neil and Buzz monitor the burn, but Mike is DISTRACTED by --

>   MIKE COLLINS (INTO COMMS)
>   Flashes out window five. I'm not
>   sure whether that's -- could be
>   something to do with the engine...

The engine?

> CAPCOM (COMMS)
> *Apollo 11, this is Houston. At 1*
> *minute, trajectory and guidance look*
> *good, and the stage is good. Over.*

> NEIL (INTO COMMS)
> Apollo 11. Roger.

Buzz and Neil look out the window, Neil UNFAZED.

> MIKE COLLINS (INTO COMMS)   NEIL (INTO COMMS)
> Watch window 5 for a second.   I don't see anything.
> See it?

> BUZZ (INTO COMMS)
> Yes, yes. Damn, everything's-- just
> kind of sparks flying out there.

Neil looks again, sees **SPARKS**. He ignores them.

> NEIL (INTO COMMS)
> ...yaw, Michael.

> MIKE COLLINS (INTO COMMS)   NEIL (INTO COMMS)
> (off Neil)   Yes, we better do that.
> ...do that?

Mike reacts to the SLIGHT PUSH and makes the adjustment as...

> BUZZ (INTO COMMS)
> Okay, 6, about 5 seconds to nominal.

POP! They're **HURLED** against harnesses into **MICROGRAVITY** again.

> NEIL (INTO COMMS)
> We have cutoff.

A notebook **FLOATS** across the cabin. He reaches for it, spots
the VISTA in the window...

NOTHING BUT BLACK. *There are NO STARS in cislunar space.*
Earth is far behind already, they're on a dark and lonely sea.
Mike stares. It's **AWESOME**. And **TERRIFYING**.

> NEIL (INTO COMMS)
> The Delta-V on the EMS: 3.3.   MIKE COLLINS (INTO COMMS)
>   Function off.

> NEIL (INTO COMMS)
> Okay, Houston, you read 11?

A beat. No answer from Houston.

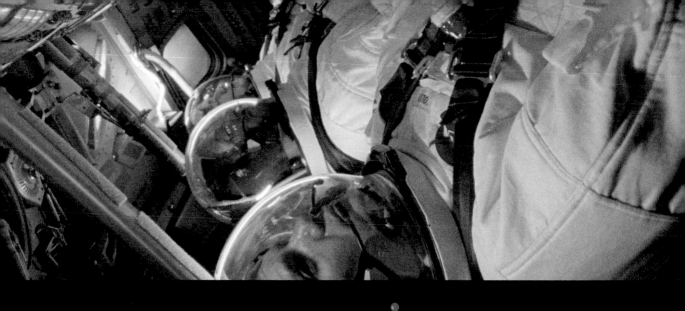

"ALMOST ALL OF THE EFFECTS ARE DONE IN IN REAL TIME OR PRACTICAL WAYS. THE ARCHIVAL FOOTAGE IS REAL FOOTAGE PROJECTED. THE VARIOUS PHYSICAL EFFECTS, THE FLYING, THE WEIGHTLESSNESS, THAT'S ALL BEING DONE IN FRONT OF CAMERA IN REAL TIME."

ADAM MERIMS, EXECUTIVE PRODUCER

                    NEIL
Not getting any answer.   MIKE COLLINS
                                        Okay, let's go to IU ACCEPT
                                        here. Why don't you try to
                                        get up high...

Buzz is concerned about silence from H   ll isn't.

                    NEIL
SCS TVC SERVO POWER 1, OFF.
        (to Buzz)
You want to get Houston on the radio
if you can?

                    BUZZ
Yes.
        (then, into comms)
Houston? Do you read?

Still nothing. Buzz is a bit UNEASY, maybe understanding for
the first time **just how alone they will be on this journey.**

                    BUZZ
Houston, do you read?

                    CAPCOM (COMMS)
This is Houston, we copy. Looks
like you're well on your way now.

                    NEIL (INTO COMMS)
Okay, Houston; we're about to SEP.
        (then, to Mike)
Mike, it's your ship now.

                                                TIME CUT TO --

A160 **OMITTED**   A160

160   **INT./EXT. APOLLO 11, TRANSLUNAR SPACE - NIGHT/DAY**   160

The SEP explosion **RIPS OFF** four panels connecting the Command
Service Module (**CSM**) to the Saturn's third stage.

The panels, **GLINTING** in the sun, float off as the CSM and
what's left of the Saturn races towards the Moon.

The quiet is **UNNERVING**, 'til RCS jets **FIRE in four directions**.

We see the CSM SLOWLY **ROTATING**, nosing towards what's left of
the Saturn... and the **LUNAR MODULE** (**LM**) within.

Sunlight cuts in and flares, shifting us in and out of day...
It's a **STUNNING BALLET**, the CSM shifting, INCHING forward...

A small PROBE extends from the CSM's nose, **SCRAPING** the LM...
NAILS on a chalkboard. *Is something wrong?*

The LM's **DROGUE** latches on. Both crafts **SHAKE**.

The CSM gently **PULLS** the LM free of the Saturn, its **FAMOUS
ORANGE HULL** SHIMMERING in the sunlight, the fragile surface
RIPPLING as it's pulled along. It's **BREATHTAKING**.

> CAPCOM (COMMS)
> *You can start PTC at your
> convenience.*

Slowly, the COMBINED CRAFT starts **ROTATING**, speeding **BACKWARDS**
towards the Moon. We watch for a moment, then **CUT INTO --**

161   **INT. APOLLO 11 CSM, TRANSLUNAR SPACE - DAY/NIGHT**   161

Buzz finishes buttoning up his flight suit as Neil, already
dressed, stows his spacesuit.

Mike, in his flight suit, PULLS OPEN a hatch and surveys the
probe and drogue connecting the LM and the CSM up close.

Mike's face **CONTORTS**.

> MIKE COLLINS
> Smells funny. Like charred
> electrical insulation.

ON NEIL, registering this. Not anxious. But it sits on him.

> MIKE COLLINS
> Wires I can see all look brand new.
> Might just be rocket fumes.

Neil remains unfazed but Mike, a bit unsettled, looks back at
Earth, **SWIFTLY RECEDING** into the distance. Buzz feels it to.

> BUZZ
> Did you bring any music?

> MIKE COLLINS
> No.

> NEIL
> Here, Buzz.

Neil who grabs a **CASSETTE TAPE** and floats it to Buzz. Buzz,
surprised, puts it in the tape deck. A familiar track, a
THEREMIN playing "Lunar Rhapsody". The odd music **SWIRLS**...

Buzz glances at Neil, working. Mike smiles.

First Man   POST CONFORMED BLUE   101.

>   MIKE COLLINS (INTO COMMS)
>   Hey, Houston, are you hearing this?

PUSH IN on Neil. **CALM.** Hearing the music... and falling into an almost **SPIRITUAL** state. **CUT TO --**

162   **EXT. TRANSLUNAR SPACE - SAME TIME**   162

WIDE ON the **TINY CRAFT**, the sun flaring the opposite side of the craft as it drifts away from us, hurtling through **INFINITE BLACK**, SILENT save for the CASSETTE PLAYER bleating out now TINNY MUSIC. It's HUMBLING.

                                                    TIME CUT TO --

163-164   **OMITTED**   163-164

165   **MISSION DAY FOUR - DAY (82:55:30)**   165

PAN OVER fogged up windows, streaked with condensation, a layer of grime over the instruments... and the men, who haven't washed in days and show varying degrees of stubble.

>   CAPCOM (COMMS)
>   *Apollo 11, this is Houston. You*
>   *should have a good view in about two*
>   *minutes, over.*

Buzz and Neil in flight suits hover over a **TOPOGRAPHICAL MAP** of the Moon.

>   CAPCOM (COMMS)
>   *When you have a free minute, could*
>   *you give us your onboard readout of*
>   *N2 Bravo, please?*

Neil continues to stare down at the map as Buzz checks nitrogen tank pressure.

>   BUZZ (INTO COMMS)
>   Nitrogen tank pressure and the tank
>   Bravo are showing 1960, something
>   like that.

>   CAPCOM (COMMS)
>   *Roger.*

Mike shoots Buzz a look. *This a problem?* Before Buzz can answer, the craft falls into shadow, the cabin turning dark...

...as slowly, something ominous FILLS the hatch window.

A **MASSIVE DARK OBJECT**...

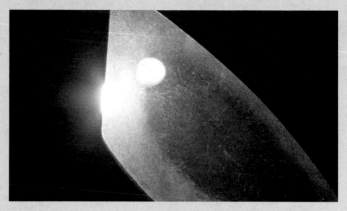

JIM: The crew had maps of the lunar surface that had been prepared by the U.S. Army Topographic Command specifically for the first Moon landing. Designed to be small enough to be handled on board, each map was 11 ½ by 10 inches and placed in a three-ring binder along with the flight plan.

JOSH: This was just one of the many props that we tried to get just right. Of course, my favorites were the sunglasses and the watches. The sunglasses were AO—they still make the original aviators and they look just as good as they did back then. And the watches were Omega. Omega actually beat out Rolex and Breitling when NASA was looking for a space watch and the good folks at Omega were kind enough to make us

twenty-two vintage watches for our astronauts to wear. And these weren't all the same—they made watches specific to three different time periods we cover in the movie.

JIM: This first close-up glimpse of the Moon was spectacular. With the Sun directly behind and hence backlighting the Moon, Apollo 11 was effectively flying through a giant solar eclipse. So what they saw at first was a massive lunar shadow, a huge dark object filling their windows, with the Sun's corona cascading brilliantly around the edges. The earthshine from behind them cast the lunar surface as three-dimensional. Mike Collins himself later said that it came as a "shock" to actually see the Moon firsthand.

JOSH: For the sake of time, we've skipped over Lunar Orbit Insertion and a night of sleep. As I mentioned earlier, we shot a scene covering the latter, but in editing, we felt the need to drive to this moment.

JIM: Mike's colorful language in Scene 168 is pretty much verbatim from the transcript, isn't that right?

JOSH: Yeah, it makes for some entertaining reading. According to Mike, he was hustling through a bunch of tasks, "like a mother hustling to get her children off on schedule to meet the school bus, and ready to relax over a cup of tea."

JIM: The last bit of dialogue between Neil and Mike (at the top of script page 103) is a fiction.

JOSH: It is. The lunar landing was the only part of the mission that hadn't been tested on Apollo 10. Neil would be flying a vehicle he'd never flown, landing on a surface no one had ever landed on—in lunar gravity, no less. It was the most dangerous part of the mission and we wanted to emphasize that. So, we played Mike's hustling as at least somewhat fueled by nervous energy and created those last two lines to bring that home.

BELOW: The Apollo 11 model shot in front of a LED screen.

"MY JOB WAS TRYING TO RECREATE WHAT IT FELT LIKE BEING TRAPPED IN THE LUNAR LANDER… AND I SAY TRAPPED BECAUSE IT'S TINY, IT'S A SARDINE CAN."

NATHAN CROWLEY, PRODUCTION DESIGNER

First Man   POST CONFORMED BLUE   102.

It's our first close view of **THE MOON**. It hovers in view, its surface **STARK** and **FOREBODING**.

The Moon is no longer a sphere in the sky but a **FULL-SCALE PLANETARY BODY**, both viscerally REAL and unbelievably **UNREAL**.

All three men stare out at the Moon, in **AWE**.

We **PUSH IN** on Neil's eyes. A bit overwhelmed by the magnitude of it all. And yet, **RESOLVED**.

166-1**OMITTED**   166-167

168   **INT. APOLLO 11, CSM, LUNAR ORBIT - MISSION DAY 5 (96:03:00)**   168

Neil, in his spacesuit, holding his helmet, prepares to enter the LM. Mike, **HUSTLING**, readies for separation, a flight manual on a lanyard round his neck.

                    MIKE COLLINS
          This is a damn three-ring circus. I
          got a fuel cell purge in progress,
          I'm watching an AUTO maneuver and --

An ALARM goes off.

                    MIKE COLLINS
          Jesus Christ.
            (turns off alarm, into comms)
          NORMAL, NORMAL.
            (into comms)
          Houston, stand by for auto alarm.
            (to Neil)
          Neil, the voice tape recorder, you
          know where that is?

                    NEIL
          Uh, no...

Neil tosses his helmet through the hatch.

                    MIKE COLLINS
          All this food and stuff up here, you
          want any of that?

                    NEIL
          No.

                    MIKE COLLINS
          Okay. Chewing gum, you want any of
          that?

> *NEIL
> Mike.

Neil looks at Mike. Mike tries to SETTLE.

> *MIKE COLLINS
> Come back, will you?

Neil nods, FLOATS TO the open hatch at the TOP of the CSM.
<u>STAY WITH NEIL</u> as he **FLOATS THROUGH THE HATCH INTO --**

169   **<u>INT. CONNECTING TUNNEL - CONTINUOUS</u>**   169

From Neil's POV, Buzz STANDS UPSIDE DOWN in the Lunar Module.
It's DISORIENTING.

> CAPCOM (COMMS)
> *Eagle, this is Houston. We see the*
> *optics zero switch on. Before you*
> *take some marks, don't forget to*
> *cycle it back off and on, and then*
> *on. Over.*

Neil BLINKS away the vertigo, pulls himself into --

170   **<u>INT. LUNAR MODULE (EAGLE) - CONTINUOUS</u>**   170

Follow Neil in, **ROTATING** with him until... <u>what was just</u>
<u>upside down is RIGHT SIDE UP</u>. He looks back Mike, *now upside*
down in the CSM, CLOSING his side of the hatch.

Neil CLOSES the LM side of the hatch.

> CAPCOM (COMMS)
> *Eagle, Houston. Could you give us a*
> *hack on the time that you switched*
> *to LM power and also verify that*
> *we're on Glycol Pump 1, over?*

Neil checks the console.

> NEIL (INTO COMMS)
> This is Eagle, we're on Pump   CAPCOM (COMMS)
> 1.   *Roger.*

CLOSE ON NEIL'S BOOTS as he plants them firmly on the Velcro
floor. He wraps a **HARNESS** around his waist, puts on his
HELMET and snaps on his **INNER GLOVES**.

> COLLINS (COMMS)
> *Eagle, Columbia. All 12 docking*
> *latches are cocked. And I'm ready*
> *to button up the hatch.*

JIM: You take some license with Neil's helmet and gloves here.

JOSH: Yes, Neil was fully suited when he entered the LM. But this is one of the few times we get to remind the audience we're in a weightless environment, so we had Neil float his helmet and gloves through the hatch.

JIM: It is true that Buzz entered the LM first (wearing only underwear) to make some initial checks, but he returned to the CSM navigation bay after Neil entered. Then he suited up and reentered the LM—at which point he and Neil sealed their side of the hatch together.

JOSH: As I mentioned, we loved the dialogue between Mike and Neil before Neil enters the LM. We were determined to include that and then follow Neil into the LM. To then show Buzz leaving the LM then coming back

after would have killed the dramatic momentum.

JIM: An hour and forty minutes. Obviously, that all happens a lot faster in the movie!

JOSH: Production Designer Nathan Crowley and his team did work very hard to get the specifics of the cabin right, which could be challenging at times. After all, while most of the Apollo 11 mission is well documented, it did happen fifty years ago, so there are occasional gaps or contradictory evidence. For example, we wound up spending some time researching exactly how Neil and Buzz tethered themselves to the LM.

JIM: Yes, I must have talked to three or four experts the week we shot that. We finally determined that it was Velcro strips plus a harness.

We hear the hatch buttoned up. Then...

                        BUZZ (INTO COMMS)
               Hey, Mike. Have you got to the
               tunnel vent step yet?

          MIKE COLLINS (COMMS)
      I'm just coming to that.   BUZZ (INTO COMMS)
                             Well, we're waiting on you.

                    MIKE COLLINS (COMMS)
               I'm ready to go to LM tunnel vent.

                    BUZZ (INTO COMMS)
               Roger. Understand.

                    MIKE COLLINS (COMMS)
               I'm going to start a maneuver now to
               our undocking attitude.

Buzz looks to Neil. Neil nods.

                    BUZZ (INTO COMMS)
               Okay.

We hear the CSM THRUSTERS repositioning the combined craft...
the Moon **TURNING SLOWLY** in the window.

                    MIKE COLLINS (COMMS)
               How about using, as an undocking
               time, 100 hours and 12 minutes?

                    NEIL (INTO COMMS)
               What have you got for AOS?

                    MIKE COLLINS (COMMS)
               I have 100 hours and 16 minutes.

                    CAPCOM (COMMS)
               Apollo 11, Houston. We are go for
               undocking. Over.

PUSH IN on Buzz, READYING HIMSELF and the cabin for the final
step of their journey to the Moon. He glances at Neil...

...who reaches for the throttle. As we **PUSH IN** on Neil, we
see the ADRENALINE FLARING...

                    MIKE COLLINS (COMMS)
               15 seconds.

A beat, then at last we HEAR the probe above them **RETRACT**.

                    MIKE COLLINS (COMMS)
               Okay, there you go.

Neil **PUSHES** the thruster. The LM slowly moves from the CSM,
sunlight moving across the windows as Neil looks towards the
waiting Moon. **PUSH OUT INTO --**

171   <u>**EXT. LUNAR MODULE, LUNAR ORBIT - DAY**</u> **(100:17:51)**   171

WE see the LM slowly PULLING BACK AWAY from the CSM...

                         CAPCOM (COMMS)
                    Eagle, Houston. We - Houston. We
                    see you on the steerable. Over.

               NEIL (COMMS)
          Roger. Eagle is undocked.   CAPCOM (COMMS)
                                              Roger. How does it look,
                                              Neil?

                         NEIL (COMMS)
                    The Eagle has wings.

ANGLE ON the CSM, receding...

A172   <u>**INT. COMMAND MODULE - SAME TIME**</u>   A172

THROUGH THE WINDOW, we see the LM. WE ROTATE, so the Moon is
below us...

...then gas jets **ROTATE** the LM, pointing the landing gear
forward, the front facing straight up, away from the Moon.

B172   <u>**INT./EXT. LUNAR MODULE, LUNAR ORBIT - SAME TIME**</u>   B172

The LM starts to drop towards the surface.

Out the window, we see the Command Module lift away as we're
plunged into inky darkness, the cabin black, save for the
console, then emergency lights...

...and then, slowly, the dark gray <u>surface of the Moon</u> rolls
backwards from the bottom of the window to the top.

C172   <u>**INT. COMMAND MODULE - SAME TIME**</u> **(102:26:28)**   C172

THROUGH THE WINDOW, we see the LM, dropping down, further and
further away, closer to the surface.

The LM begins orbiting around the Moon, heading into night.

JOSH: We take some visual license with the landing. In reality, the crew entered daylight just before undocking and remained in daylight until just before landing. They had been flying in and out of darkness on every orbit and we wanted to get this dynamic across.

JIM: The comms here and throughout the landing are edited down, but the actual lines are verbatim or close to it.

JOSH: Yes. Occasionally we add a line or two of dialogue to help the audience understand what's going on and what the potential issues are. But the actual transcript is thrilling, so we tried to stick to it as much as possible. Again, the 'Apollo Flight Journal' was helpful here. And there's a great site (www.firstmenonthemoon.com) that runs through the last fifteen minutes of the landing with video, comms, and the flight director loop all playing at once. That was another good resource.

172 <u>**INT. LUNAR MODULE - SAME TIME** (102:26:55)</u>   172

A **500 CODE** pops up on Buzz's console. Buzz moves to descent 1
then back to auto. The alarm goes away.

Buzz notices Neil **STRUGGLING** with a LOOSE BREAKER.

                    NEIL
          It just won't stay...

                    BUZZ
          We'll have to tell them about that.

CLICK. Neil **FIRMLY PUSHES** the breaker back in. <u>It stays</u>.

                    NEIL
          Let's prep for descent.

Neil gets to work, like nothing happened. For a moment, Buzz
pauses, <u>knowing his life's in Neil's hands</u>.

     CAPCOM (COMMS)
Eagle, you're go for powered   BUZZ (INTO COMMS)
descent.   Roger, we read you.

Buzz gets back to it as the sun breaks. He and Neil can see
the lunar surface is even closer...

Two indicator lights on the DSKY flash on.

                    BUZZ
          Altitude light's on, we don't have
          radar data.

                    NEIL
          Let's proceed.

Buzz eyes Neil. He eyes the clock. No time to disagree.

     BUZZ (INTO COMMS)
Proceed. 1, 0...   NEIL (INTO COMMS)
                                            Ignition.

A173 <u>**SMASH OUTSIDE THE CRAFT --**</u>   A173

The thruster comes to life, thrusting into the empty void.

B173 <u>**INSIDE THE LM**</u> (102:35:14)   B173
Neil eases the throttle forward and the lunar module **LURCHES**
ahead... towards the surface.

Neil eyes the radar data, puts RADAR MODE SWITCH in SLEW, then
looks to the window, **WATCHES** the landscape pass the hashmarks.

Buzz sees Neil's eyes ticking from the window the clock.

> NEIL (INTO COMMS)
> ...went by the 3 minute point early.
> Our position checks downrange show
> us to be a little off...

> CAPCOM (COMMS)
> *Roger. Co...*

HEAVY STATIC again. Shit. Then...

> CAPCOM (COMMS)
> *...go to... You are go to*
> *continue powered descent.*

> BUZZ (INTO COMMS)
> Roger.

A bit of a relief, until a button on the console **FLASHES ON.**

> NEIL (INTO COMMS)
> Program alarm.

1202. It beeps.

BUZZ	NEIL
What's a 1202?	I don't know.
	(into comms)
	Houston, give us a reading on
	the 1202 program alarm.

The beeping seems to grow **LOUDER.** Neil's eyes **TICK RAPIDLY**
from the alarm to his DROPPING ALTITUDE GAUGE...

A **TENSE BEAT,** then --

> CAPCOM (COMMS)
> *Roger, we got-- we're go on that*
> *alarm.*

Buzz shuts it off. Neil checks his instruments.

> BUZZ
> Looks like about 820...

Another button **FLASHES ON.** The same **BEEPING** again...

> BUZZ
> Same alarm.

Neil reacts, uncharacteristically **ANNOYED.** Buzz shuts it off
as Neil **YAWS** the craft. The Moon **SLIDES** out of the window...

> CAPCOM (COMMS)
> Roger. We're go on that alarm.

First Man   POST CONFORMED BLUE   108.

We now see nothing but space. Neil continues to pitch.

The LM pushes over, feet towards the surface and Buzz sees THE MOON again, the surface now **RISING UP** to meet them.

                    NEIL
          Okay. 3000 at 70.

A **THIRD ALARM** beeps. A new one. This time Neil **IGNORES** it.

        BUZZ (INTO COMMS)
Program alarm, 1201.   CAPCOM (COMMS)
                                    Roger, 1201 alarm.

Buzz looks to Neil, but Neil keeps flying.

                    NEIL
          2000 at 50.

Buzz stares at the alarm, beeping **PERVADING** the tiny space.

                    CAPCOM (COMMS)
          We're go. Same type. We're go.

Neil PUNCHES OFF the alarm, **IRRITATED**. Buzz glances over at him, sensing the determination. **CLOSE ON** Neil's eyes, then --

C173 **SMASH OUTSIDE THE CRAFT --**   C173

Our biggest wide yet, the tiny fragile LM hurtling over the vast lunar surface, as ominous a landing pad as one has ever seen. We hold on this a moment, then we...

D173 **SMASH BACK INTO THE LM** (102:42:32)   D173
Neil looks out the window, getting a first look at the landing area. Neil's eyes **NARROW**.

There's a HUGE CRATER, A HUNDRED YARDS ACROSS.

                    NEIL
          Give me an LPD angle.

                    BUZZ
          47 degrees.

Neil eyeballs it. It's okay. They'll be short of the crater.

                    NEIL
          Okay, we'll be short of that crater.

He continues to descend. Eyes **FIXED** on the landing area.

JIM: Some explanation here – Neil is tracking his surface landmarks (which he'd studied on the lunar maps) against time of PDI (Initiation of Powered Decent) to confirm Eagle's trajectory. In this moment, he sees Eagle is passing over the Maskaleyne W crater a few seconds early, which tells him the descent trajectory is taking them a little long.

Neil could not be sure why they were over the crater early, but he accurately guessed that PDI must have started a little late. This was likely because incomplete venting had led to residual pressure in the tunnel between the LM and the CSM, which in turn had given Eagle a little extra 'kick' on undocking. The result? A velocity-induced positional error that put Eagle a good distance away from where it was supposed to be. In subsequent Apollo flights, Mission Control made sure to double-check the status of the tunnel pressure before approving the LM's undocking.

JOSH: This 1202 alarm must have been somewhat disconcerting.

JIM: Yes. It was the first of several program alarms that went off during the last phase of the Eagle's descent. There is no question that the alarms and attendant lights distracted Neil and Buzz and added drama to the last minutes before landing.

JOSH: What fascinates me is that neither Neil nor Buzz knew which of the dozens of potential alarms 1202 represented. For a good twenty seconds, they had no idea how big an issue this was.

JIM: The 1202 alarm was triggered by an overload in the on-board computer caused by the inflow of just-arriving landing radar data. And it was inconsequential. But Neil and Buzz did not know this, as they had not simulated program alarms in training.

What's amazing is that, eleven days prior to the Apollo 11 launch, Richard Koos, the Sim Supe at the Manned Spacecraft Center had put Gene Kranz's Mission Control White Team

through exactly this simulation. Koos put a 1201 alarm to Kranz's team and suddenly Steve Bales, the young LM computer system expert, was on the hot seat. At that point, Bales had no mission rules on program alarms. And although everything seemed to be working, the alarm said the computer was overloaded. Unable to diagnose the cause (or the ramifications), Bales called for an abort. Sim Supe Koos was not happy, "If the guidance was working, the control jets firing, and the crew displays updating, then all mission critical tasks were getting done. This was not an abort."

After this, Koos put Kranz's team through four additional hours of training on program alarms—which led Bales to create Mission Rule 5-90, Item 11, specifying what program alarms required an abort (1201 and 1202 were not among them). So, when the alarm came up for Neil and Buzz and GUIDO Jack Garman brought the alarm code to the attention of Bales and his team of LM computer experts in the back room, Bales was able to quickly relay to Kranz, "We're go on that alarm." The same held for the other alarms.

JOSH: It's a great story. A shame we didn't have room for it.

JIM: You know, Rick Houston talked a lot about how tense it was in Mission Control at this moment (and later when Neil is low on fuel). I found it interesting that you and Damien didn't want to cut back to Mission Control.

JOSH: We wanted to remain grounded in Neil's point of view. We wanted to know only what Neil and Buzz knew so we could feel how they felt. While cutting back to Mission Control would have been the traditional move, we thought that living in the cramped Eagle cabin with moments of complete uncertainty would help the audience understand what Neil was going through at the time.

"DAMIEN WANTED TO CREATE AN ENVIRONMENT THAT WAS REALITY FOR THE ACTORS."

J.D. SCHWALM, SPECIAL EFFECTS SUPERVISOR

> BUZZ (INTO COMMS)
> 700 feet, 21 down, 33 degrees.

NEIL'S POV. _The landing area in front of the crater is not flat_. It's covered with **GIANT BOULDERS.**

> *NEIL
> Pretty rocky area.

> *BUZZ (INTO COMMS)
> (follows his gaze)
> Those boulders are as big as cars. We can't land there.

Buzz is right. Neil makes a **QUICK DECISION...**

> NEIL
> I'm going to manual...

Neil takes over manual control. We hear the **POP** and **HISS** of the Descent Propulsion System (**DPS**) as the craft **PITCHES OVER.**

> BUZZ (INTO COMMS)
> 540 feet, down at 30, down at 15.

Buzz eyes the gauges...

> BUZZ (INTO COMMS)
> 330... 300 feet, down 3 1/2.

Neil adjusts. Buzz watches the PROPULSION CONSOLE. **CLOSE ON** the FUEL DESCENT MONITOR #2. **DROPPING.** _12, 11, 10..._

> BUZZ
> Fuel's at eight percent.

Neil keeps flying, focused, **INTENT**, even as the altitude and velocity lights **FLASH** on the DSKY.

> BUZZ
> Radar's lost track with the surface
> again.

But Neil grips the throttle, eyes **TICKING FURIOUSLY** from the window to the ALTITUDE GAUGE. As the DPS **HISSES...**

> BUZZ (INTO COMMS)
> 160 feet, 6 1/2 down... 5 1/2 down,
> 9 forward.

RACK TO Buzz, his eyes ticking from the ALTITUDE GAUGE to the **PROPULSION CONSOLE. CLOSE ON** the fuel numbers. _8, 7, 6..._

> _CONTROL (COMMS)_
> _Low level._

> _FLIGHT (COMMS)_
> _Low level._

JOSH: Now we arrive at the second major hurdle Neil and Buzz faced. Neil sees a pretty big crater (West Crater) and wants to make sure they're not heading right into it, so he asks for the LPD angle.

JIM: LPD stands for Landing Point Designator. Neil is looking through a set of hash marks on his window and the LPD angle, which Buzz gives him from the Primary Guidance and Navigation System (PGNS), will tell Neil where to look along the vertical scale to find the place the computer thinks they will land.

JOSH: Neil can see they're going to be short of the crater. Which would've been fine... but when they get a little lower, Neil sees that area is covered with boulders "the size of Volkswagens." We added Buzz's line (highlighted) so the audience will understand they can't land there; but Neil's "pretty rocky area" is verbatim.

JIM: This is why Neil takes over manually. As he approaches 500 feet, he tips the vehicle over to roughly zero pitch, slowing the descent. By pitching nearly upright, he also maintains his forward speed—some 50 to 60 feet per second—so that, like a helicopter pilot, he could fly beyond the crater.

JOSH: But now he's got another problem. Fuel.

JIM: Yes. And now everyone in Mission Control starts holding their breath. They didn't know why Neil had switched to manual; they couldn't see what Neil saw. But what they could see is his fuel supply was getting preciously low. As I explain in my book, the "low level" call means that the propellant in the tanks of the Eagle had fallen below the point where it could be measured, like a gas gauge in an automobile showing empty with the car still running. As Gene Kranz said, "I never dreamed we would still be flying this close to empty."

JOSH: One small fudge here—Neil and Buzz did not hear the flight director loop, they only heard CapCom Charlie Duke. But we include some of the flight director loop (marked as V.O.) to help the audience understand they're in some trouble. And it got worse. When the Quantity Light flashed, that meant they were under 5% of the original fuel load. This triggered the Bingo countdown.

JIM: A "Bingo" call would mean "land in 20 seconds or abort." If the Bingo countdown got to zero, Neil would have 20 seconds to land; otherwise, the Mission Rule dictated that he should abort immediately.

JOSH: In reality, all Buzz said here was "Quantity Light" as Neil knew exactly what that meant. But to help the audience we have Buzz give us some detail—94 seconds to Bingo, so we're 114 seconds to abort.

JIM: Before Neil and Buzz land they'll get down to 20 seconds and 2% fuel supply. No wonder Charlie Duke got a little tongue-tied just after the landing!

JOSH: That's actually a funny story. We sent the script to Charlie for notes and he was quite helpful. But it led to the most embarrassing note I've ever gotten—I'd written his iconic line as, "You got a *lotta* guys about to turn blue." Which is why I like to run these scripts by the experts!

>   BUZZ (INTO COMMS)
> 120 feet. 5 percent fuel remaining.

Buzz glances at Neil, **CLOCKS** his intensity... and at that moment he knows. **Neil's landing this ship.**

The FUEL QUANTITY LIGHT **FLASHES ON.**

>   *BUZZ
> Quantity light. 94 seconds to
> bingo, 114 to mandatory abort.

Buzz clocks the **FIRE** in Neil's eyes, DETERMINATION. <u>Jesus.</u>

>   BUZZ (INTO COMMS)
> Down a half, 6 forward.

Neil's eyes **TICK** from the gauges to the window...

> CONTROL (COMMS)   FLIGHT (COMMS)
> Standby for 60.   Rog.

>   CAPCOM (COMMS)
> 60 seconds.

...but it's unclear if Neil's even <u>LISTENING anymore.</u> We push in on Neil as Buzz continues to call the descent...

Buzz looks at the fuel. **CLOSE ON** the numbers. *3, 2...* **FUCK.** <u>Buzz's eyes **WIDEN**.</u>

>   BUZZ (INTO COMMS)
> 40 feet, down 2 1/2...

> CONTROL (V.O., COMMS)
> Standby for 30.   CAPCOM (COMMS)
>   30 seconds.

>   BUZZ
> 20 feet, down a half. Drifting
> forward just a little bit.

Buzz eyes the fuel gauge, approaching **ZERO...**

...then, out the window, **DUST SWIRLS** up from the lunar surface. And, before they realize what's happening...

A **BLUE LIGHT FLASHES** on the console. Buzz blinks in **DISBELIEF**. [MET is *102:45:40*].

>   BUZZ (INTO COMMS)
> <u>Contact light.</u>
>   NEIL (INTO COMMS)
>   Shutdown.

Neil, <u>SPENT</u>, lets the throttle SLIP from his hands. He flips the switch. The dust settles and we see... **the LUNAR SURFACE stretching out in front of them.**

Neil stares, BLANK. Buzz is STUNNED. A beat, then Buzz recovers, starts powering down the LM.

> CAPCOM (COMMS)
> *...we copy you down, Eagle.*

> NEIL (INTO COMMS)
> Houston, Tranquility Base here. <u>The Eagle has landed</u>.

An **ENORMOUS CHEER** goes up in Mission Control. Neil **FLINCHES**. It's VISCERAL; <u>something about it hits him right in the gut</u>.

> *****CAPCOM (COMMS)
> *Roger, Twan, Tranquility, we copy*
> *you on the ground. You got a bunch*
> *of guys about to turn blue. We're*
> *breathing again, thanks a lot.*

> NEIL (INTO COMMS)
> Thank you.

Neil sits, <u>struggling to process</u>. Buzz extends a hand.

> BUZZ
> Very smooth touchdown.

Neil nods. **PUSH IN** on his eyes, deep PAIN and JOY battling within as we **MATCH CUT TO --**

173 **INT. LUNAR MODULE, LUNAR SURFACE - LATER (108:20:00)**   173

An **EVA BOOT** is pulled on over an under boot. An **ANKLE STRAP** is pulled tight; top boot **VELCRO** and **BUTTONS** are closed.

A **PLSS/EMU BACKPACK** is lifted onto a torso. **METAL CLIPS** at the waist and chest are fastened to hold the pack into place.

O2/CO2 **HOSE NOZZLES** (**BLUE/RED**) are plugged into suit **PORTS**.

An **RCU** is clipped into place with **METAL CLIPS**.

A PLSS **HOSE** and **CONNECTOR** is plugged into the RCU.

An **OXYGEN DIAL** is turned up on the RCU. We hear a familiar **HISS** as we **REVEAL --**

Neil. Bubble helmet, PLSS, EMU and RCU on. Hoses connected.

Neil and Buzz don inner gloves, checking wrist locks. Meticulous, cognizant of the DANGERS on the surface.

Neil uses the **MIRROR** on his wrist to look at the controls on the RCU. He flips a switch. We hear water **WHOOSH** as it circulates through Neil's LCG.

And now we hear a **HISS** as they pressurize the suits. A beat.

Buzz opens a valve. We **HEAR** AIR VENTING OUT. It's nerve-wracking, all that air disappearing into space.

Buzz kneels, grabs the hatch handle, rotates it counter-clockwise and pulls... but the hatch won't budge.

Buzz pulls again. Struggling. He stops, FRUSTRATED.

Neil looks at Buzz... then Buzz tries again. This time, the door **PULLS OPEN**... revealing the barren surface below.

All sound is sucked out and **EVERYTHING GOES SILENT**.

Neil and Buzz **STARE**, taking in **THE WORLD OUTSIDE THE DOOR**. In this moment, we understand how odd, how **UNDENIABLY STRANGE** it is to be parked on the surface of another heavenly body.

A beat. Neil steps forward, puts his gold visor down, and turns around to start backing out. **TIME CUT TO --**

174   **EXT. LUNAR MODULE, LUNAR SURFACE** (109:21:09)   174

TIGHT ON Neil, gold visor down now, as he SHIMMIES back through the hatch onto the porch, Buzz helping guide him, holding tight the LEC that serves as a safety tether.

Neil steps down onto the top rung of the ladder.

Neil pulls the D-ring which releases the **MESA,** attached to the side of the LM under Buzz's station.

The MESA swings down into position... and we spot the **CAMERA** now pointed at Neil.

Neil continues to move down the ladder. He stays focused, eyes on the LADDER, HANDS, FEET...

> CAPCOM (COMMS)
> *Okay, Neil, we can see you coming*
> *down the ladder now.*

Neil just continues moving down, reaching the final rung... then hopping off of it to the footpad of the LM. But before turning around, he quickly jumps back up to the first rung.

"YOU COULD STAND OFF THERE AND WATCH, SEE THE LUNAR MODULE, SEE THE GUYS COME DOWN THE LADDER AND YOU'D SWEAR YOU'RE STANDING ON THE MOON."

ALFRED WORDEN, COMMAND MODULE PILOT FOR THE APOLLO 15

JOSH: Neil and Buzz had a fair amount to do after landing and the flight plan called for a four hour rest period. Not surprisingly, Neil and Buzz decided to proceed with the EVA right away. Still, Neil didn't set foot on the Moon until seven hours after landing.

JIM: Opening the hatch of the LM proved to be quite a chore. Even when the astronauts got the cabin pressure down to a pretty low psi, it took 200 pounds of pressure to open. That was more pressure than Buzz could easily manage (mostly it was Buzz doing the pulling because the door opened in his direction). And obviously, Buzz didn't want to bend or break anything! So he waited until the pressure got even lower—even then he had to try a number of times to open it.

JOSH: Just as the 'Apollo Flight Journal' was useful when studying the flight to the Moon, so the 'Apollo Lunar Surface Journal' was useful when studying the EVA out onto the surface. Of course, Frank Hughes provided us with an original Apollo 11 Lunar Surface Operations Plan from June 27, 1969. So that was pretty helpful as well.

JIM: A good example is the tether Neil wears as he goes down to the surface. It wasn't until 2004 that a few folks began to realize Neil might have used the Lunar Equipment Conveyor (LEC) as a safety tether when he walked down the ladder to the surface of the Moon. Neil hadn't recalled that himself, but the folks working on the 'Apollo Lunar Surface Journal' concluded that he had used the LEC this way. Frank's Operations Plan confirmed this as well, specifying that Buzz should "play out the LEC and use as a safety tether."

JOSH: We also wanted to make sure to include shots of the MESA. Conspiracy theorists have long asked, "How did they get that shot of Neil coming down the ladder?" The MESA is the answer.

JIM: The MESA (or Modularized Equipment Stowage Assembly) holding the Apollo TV camera was stored under the LM. There was a D-ring release at the top of the ladder—Neil pulled it and the MESA swung down into position, along with the camera. The camera had to be stowed upside down on its top (as that was the camera's only flat surface), so the image was initially upside down; they had to flip it in the studio back on Earth.

>   NEIL (INTO COMMS)
>   Okay. I just checked getting back
>   up to that first step, Buzz. It's...
>   the strut isn't collapsed too far,
>   but it's adequate to get back up.

>   CAPCOM (COMMS)   NEIL (INTO COMMS)
>   *Roger, we copy.*   Takes a pretty good little
>   jump.

A beat, then Neil hops down to the LM footpad again.

>   CAPCOM (COMMS)
>   *Buzz, this is Houston. F/2 -*
>   *1/160th a second for shadow*
>   *photography on the sequence camera.*

Neil pauses, staring out. A million things going through his
mind, the **ODDNESS** of it all, the **DESOLATE BEAUTY**... The years
of work and sacrifice. And Elliot. And Ed.

The **ENORMITY** of what he's done almost **OVERWHELMS**.

>   NEIL (INTO COMMS)
>   I'm, uh, at the foot of the ladder,
>   the LM footbeds are only uh...
>   depressed in the surface about...
>   uh, one or two inches, although the
>   surface appears to be very, very
>   fine grained as you get close to it.
>   It's almost like a powder. Down
>   there, it's very fine.
>   (then)
>   I'm gonna step off the LM now.

And now he steps down **ONTO THE MOON**... oddly **DISCONNECTED** from
the moment, the line he's prepared...

>   *NEIL (INTO COMMS)
>   That's one small step for a man, one
>   giant leap for mankind...

The COMMS **FADE**. Neil turns, taking it all in...

A beat, then he ties down his LEC to the ladder and steps
around the side of the LM. He looks up, **FLIPS OPEN** his visor.

And now he sees the **EARTH**, hanging high over the lunar
horizon. It's stunning.

Off the look in his eyes, we **PRELAP** --

>   BUZZ **(PRELAP)**
>   Okay, ready for me to come out?

JOSH: While we didn't include all the comms here, we tried to be faithful with the comms we did include. We skip over Neil's initial egress (he left the LM then practiced getting back in) and begin with the second egress as Neil walks down the ladder. Neil gets to the bottom and jumps back up, again to practice, then gets to the foot of the ladder.

JIM: The Apollo EVA helmets had two visors—a clear internal visor and a gold tinted external visor to protect the astronauts from the harsh, direct sunlight they'd experience on the lunar surface. I consulted Apollo experts from all around the world trying to figure out if Neil's external (gold tinted) visor was up or down when he stepped onto the lunar surface. Written records didn't answer the question satisfactorily, and none of the photographic evidence provided a definitive answer.

JOSH: Given the lack of clear evidence, we decided to make a creative choice—keep the visor down until Neil's personal moment at the lip of Little West Crater in Scene 177.

JIM: Ironically, after ten days of calling every Apollo expert I could think of, I concluded that Neil probably had his gold visor up for the early part of his EVA, including the moment he stepped off the ladder.

JOSH: Speaking of that first step, Neil was always much more focused on the landing than the walking on the Moon.

JIM: Yes, the landing was the real challenge. The walking was secondary. He always said he didn't prepare what he was going to say when he stepped onto the Moon; that's probably why. Of course, Neil intended to say something different from what the world heard.

JOSH: "That's one small step for a man, one giant leap for mankind."

JIM: Yes. He might have said that. He certainly believed he did, but no one heard it that way. As Neil himself would tell me: "I think that reasonable people will realize that I didn't intentionally make an inane statement (because 'man' and 'mankind' would mean the same thing), and that certainly the 'a' was intended, because that's the only way the statement makes any sense. So I would hope that history would grant me leeway for dropping the syllable and understand that it was certainly intended, even if it wasn't said— although it actually might have been."

"YOU'RE USING ALL OF YOUR EXPERIENCE TO DO SOMETHING FOR THE FIRST TIME... YOU DON'T HAVE TO ANSWER THE QUESTIONS. YOU HAVE TO DISCOVER THEM THROUGH DESIGN."

NATHAN CROWLEY, PRODUCTION DESIGNER

JIM: So, this scene is bound to be controversial. On the one hand, Neil never said he'd taken anything of Karen's to the Moon. On the other hand, there is no definitive record of what he took with him to the lunar surface. When I asked Neil's younger sister June (Armstrong Hoffman) if she thought he took anything of Karen's to the Moon, June answered with tears in her eyes, "Oh, I dearly hope so." Same goes for me.

JOSH: Landing on the Moon was a tremendous achievement. But it also was the end of an incredibly harrowing journey. Emotionally, we wanted to get back to the origins of that journey in order to help it feel more personal, to help put the audience in Neil's shoes. This is why in post we started playing with the Juniper Hills flashback here. Of course, we'd always imagined tossing the bracelet as a moment of release.

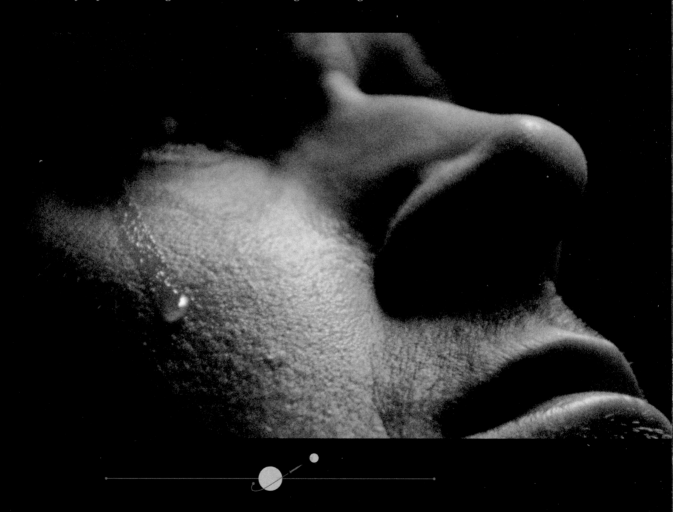

"THERE WAS A DEEP FOREBODING FOR WHAT LIFE WOULD BE LIKE FOR THESE MEN ONCE THEY RETURNED. THAT IN SOME WAY WE WOULDN'T BE ABLE TO REACH THEM BECAUSE THEY HAD BEEN SOMEWHERE THAT NONE OF US HAD BEEN."

WYCK GODFREY, PRODUCER

First Man   POST CONFORMED BLUE   114.

175   **EXT. LUNAR SURFACE - MOMENTS LATER (109:43:16)**   175

WIDE ON Neil collecting a soil sample as Buzz comes down off
the LM to the lunar surface. We **HOLD ON** Neil as --

       BUZZ (COMMS)   NEIL (INTO COMMS)
Beautiful view.   Isn't that something?
                               Magnificent sight out here.

       BUZZ (COMMS)
   Magnificent desolation.

We **CUT TO** --

176   **EXT. LUNAR SURFACE - LATER**   176

Neil alone. Slowly moving across the lunar surface. He
stops. Looks down, picks up his foot... sees his **FOOTPRINT.**

He looks up. And back behind him. Catching site of Buzz,
loping across the surface. He looks back further. Sees the
LM in the distance. Impossibly fragile. Off Neil, **CUT TO** --

177   **EXT. LUNAR SURFACE - LATER (111:11:15)**   177

The lip of LITTLE WEST CRATER. Visor now lifted, Neil stares
into the crater. Deep and vast, like nothing we've seen.

Off the **UTTER BLACKNESS** of the long shadows, we **FLASH TO** --

*Juniper hills. The cabin. Rick. Janet, younger, carefree.
And Neil. With Karen. All of them happy. Paradise lost.*

We linger for a moment, then **FLASH BACK TO** --

Neil peers into the void, holds a **FAMILIAR BRACELET.** Karen's.

PUSH IN on Neil's eyes, on the PAIN. A beat. Then he **FLICKS**
Karen's bracelet into the crater...

It flies on and on and on, falling at last into the abyss...

Neil's **UNABLE TO CONTROL HIMSELF. A TEAR FALLS,** a bevy of
emotions rising to the surface. All swirling round until...

...the **TEARS COME FREELY,** raining down with all the pent up
feelings. The first and last such outburst we'll ever see.

178   **OMITTED**   178

179   **EXT. LUNAR MODULE, MOON - DAY 6 (124:19:59)**   179

We're on the side of the LM, looking down at the Moon.

> BUZZ (COMMS)
> *...6, 5, abort stage -- engine arm,*
> *ascent, proceed...*

The thrusters **IGNITE**. DUST kicks up as the ship **ROCKS BACK AND FORTH**, lifting off the surface...

...and casting a **LONG SHADOW** as it rises.

As the ship pulls away from the Moon, we hear familiar voices from a television broadcast...

> WALTER CRONKITE (O.C.)
> *It may not be a beauty one can pass*
> *on to future beholders. These first*
> *men on the Moon can see something*
> *that men who follow will miss...*

> ERIC SEVAREID (O.C.)
> *Yes, we're always going to feel,*
> *somehow, strangers to these men...*
> *disappeared into another life that*
> *we can't follow. I wonder what*
> *their life will be like, now. The*
> *Moon treated them well. How people*
> *on earth will treat these men...*
> *that gives me more foreboding...*

Off the now distant Moon, we **SMASH TO --**

A180 **APOLLO 11 CELEBRATION MONTAGE**   A180

In a series of QUICK CUTS, we see **TV FOOTAGE** of crowds around the world watching the Apollo 11 landing; of individuals waxing on in amazement at the feat. This takes us to --

INT. ASTRONAUT LOUNGE, QUARANTINE FACILITY, MSC - JULY 28, 1969

CLOSE ON a SEA of MAGAZINES and NEWSPAPERS: **TIME, LIFE,** the **NEW YORK TIMES**... ***All with front page coverage of Apollo 11.***

FIND BUZZ hovering over it all, staring at a TV. **MESMERIZED.**

> NEWSCASTER (ON TV)
> *And in Washington, an anonymous*
> *citizen has placed a small bouquet*
> *on the grave of John F Kennedy with*
> *a note, "Mr. President, the Eagle*
> *has landed." And indeed, on this*
> *day, it's hard not to think back*
> *upon that speech our 35th President*
> *gave at Rice University just seven*
> *short years ago...*

JOSH: The Cronkite and Sevareid lines come from the from the actual CBS broadcast (and your book) and I just loved them. They seem to sum up everything about what things are going to be like for these men when they come home. The lines originally continued over a montage of re-entry and the crew's trip to the MSC Quarantine Facility. But when we got into editorial, we felt we needed more of a sense of how singular a moment this was—after all, 600 million people, one fifth of the world's population, tuned in to the landing. So we trimmed the language and replaced the montage with worldwide reactions to the landing. I think it makes the ending that much more powerful.

"WE SHOT IN IMAX TO MIMIC THE PHOTOS YOU'VE SEEN FROM THE MOON. A GREATER NEGATIVE, MUCH MORE DETAIL—WE WANTED THE AUDIENCE TO BE IMMERSED IN THIS WORLD."

LINUS SANDGREN, CINEMATOGRAPHER

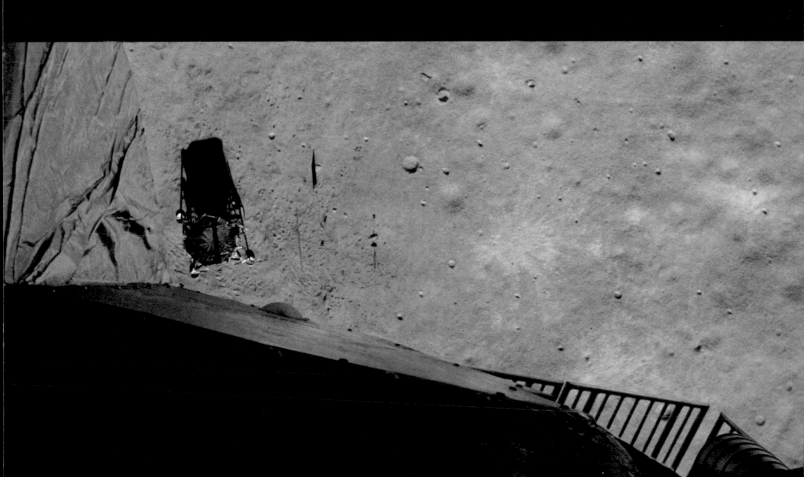

First Man   POST CONFORMED BLUE   116.

Neil walks in, glances at the TV. JFK's 1962 speech at Rice.

> PRESIDENT KENNEDY (ON TV)
> *But why, some say, the Moon? Why choose this as our goal? And they may well ask why climb the highest mountain? Why fly the Atlantic? We choose to go to the Moon. We choose to go to the Moon in this decade and do the other things, not because they are easy, but because they are hard...*

Neil eyes Buzz, who seems **BEWILDERED.** Off Neil, processing --

EXT. ARMSTRONG HOUSE - HOUSTON, TX - DAY

CLOSE ON Janet as she walks briskly across the front yard, trying not to look at the **THRONG OF REPORTERS**, the **HUGE DISPLAY** of FLOWERS and SIGNS on the lawn...

...but the reporters CORNER HER as she reaches the car.

> NBC REPORTER
> Mrs. Armstrong, have all your prayers been answered?

> JANET
> Yes, yes they have.

She **FORCES** a smile.

> REPORTERS
> How would you describe the flight?

> JANET
> I could only say that it was... out of this world.

Another forced smile and then she gets into the car.

187   **INT. HALLWAY, QUARANTINE FACILITY, MSC - LATER**   187

Deke walks Janet down a lackluster hall.

> DEKE (O.C.)
> They will be quarantined for the full three weeks, but there's no sign of infection or disease.

They reach a door. Just before he lets her in...

> DEKE
> Congratulations, Jan.

Janet walks into --

First Man   POST CONFORMED BLUE   117.

INT. PRESS ROOM, QUARANTINE FACILITY - CONTINUOUS

The NASA logo. And a GLASS WALL. On the other side of it, in a large room, Neil stands in civilian clothes.

Deke closes the door behind Janet, leaving her alone with Neil. She walks to the glass. He sits and she does as well.

A long beat.

They **STARE** at each other through the glass.

At last, Neil lifts a **HAND**... then Janet does the same.

They **PRESS HANDS TOGETHER**, from opposite sides of the glass.

And off that small ray of **HOPE**...

...that maybe they might navigate the gulf between them, that they might find a way back to each other again, we...

                                        FADE OUT

                        THE END

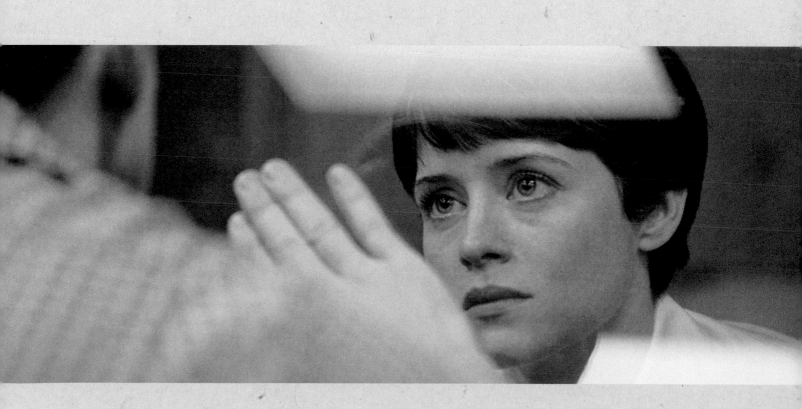

AFTERWORD

RICK AND MARK ARMSTRONG

After the book *First Man* was published in 2005, and while playing golf with my father, I asked him to summarize his thoughts on the biography that he had authorized. We were on the second tee of Camargo Golf Club in Cincinnati, Ohio, a tricky par 5 featuring a sharp dogleg to the right. His response was that the project had gone well, but there were a few cases where he didn't recall things exactly the way they were printed. As I prepared to ask for elaboration, dad hit a shot into the high grass to the right of the fairway and our attention turned to the inevitable and familiar hunt for his golf ball. It was a clear summer day, one of many (albeit too few) days we spent walking the fairways together—moments that I will always cherish. One thing led to another and we never returned to our original conversation. This is often how it was with dad. There simply never seemed to be enough time, or the right moment, for in-depth conversation. Dad was a shared global commodity, and the legitimate and worthwhile demands on his time were far more than he could ever negotiate. So, you enjoyed the time you had together, and you tried to do whatever you could to lessen the burden that was clearly and perpetually on his shoulders.

My experience with the making of the movie *First Man* has been a parallel one. The project has gone well, very well in fact, and I was pleased to have been consulted and included. And while not everything in the film transpired exactly the way it has been presented, this annotated script has given the writers an excellent vehicle for explaining what and why things were fictionalized, and, as such, serves as a primer on storytelling through film.

MARK ARMSTRONG, JULY 2018

Dad once advised me that you have to be careful what you say in public because there will always be someone ready and willing to disagree with you. However, one thing I can safely say about dad without fear of contradiction is that he was a stickler for accuracy. For example, a few months before he passed away we were in a rain delay on the golf course during a Father's Day tournament, and he was describing the events around a *60 Minutes* news segment that had attributed some comments to him that were critical of the commercial space efforts at the time— comments that he didn't actually make. Sitting under the shelter in our golf cart, he described the considerable efforts he had to go through to get a retraction/correction, which involved pursuing numerous communication channels until he got satisfaction, which he eventually did.

Having been fortunate to be involved with the *First Man* film since March of 2016, I have seen first-hand that the entire production company was committed to getting the facts and details right, even for things that will not likely be noticed by the audience. For example, the prop department did their best to track down the kind of pool float we had (a large styrofoam bowl we called the "*U.S.S Eggshell*"). That's impressive attention to detail I would say! For those interested in what is true and what isn't and why, I hope you find this book both interesting and enlightening. I think dad would have been pleased with the effort of all involved.

RICK ARMSTRONG, JULY 2018

P.S. In the movie *Apollo 13*, there is a scene where dad is at the Lovell's house and Jim Lovell's mother asks if he is in the space program too. Dad was there at the house during the flight, but in reality he already knew her— but dad said the change "made a better movie"— so I guess in certain circumstances he was willing to let accuracy slide (as he was fond of saying) "just a tad."

ACKNOWLEDGEMENTS FROM JOSH SINGER

This movie and this book would not have been possible without all the help from NASA and the folks who knew Neil best. At NASA, the odyssey began at headquarters, where Bert Ulrich and Bill Barry gave me more books than I could carry and quizzed me on Korolev. Bert and Bill have been invaluable help throughout the process. I next went to Armstrong Flight Research Center, where Cam Martin introduced me to Gene Matranga, Christian Gelzer, and others who helped me understand Neil's work for NASA FRC in the late 50s and early 60s. At JSC, Jeannie Aquino put me in touch with historians Rebecca Wright and Jennifer Ross Nazzal who sent tons of primary material. She also had us talk to such incredible sources as Glynn Lunney, Gene Kranz, Tom Sanzone, George Abbey, Frank Hughes, and Joe Engle. Those last two were crucial—as must be apparent from the book, we simply could not have made the film without Frank and Joe. Frank tells the best stories you ever heard about the program, and there's nothing quite like having Joe walk you through an X-15 flight.

Jim Hansen put us in touch with Rick and Mark Armstrong, who have continued to give their time and effort to helping us try to get things right and have shown remarkable kindness towards us on our first few drafts (when we truly were stumbling around in the dark). Their notes at script and in post have been a true compass. They also put us in touch with Neil's sister June, who spent invaluable time with Damien and Ryan. Finally, Rick and Mark were kind enough to put us directly in touch with Janet, who spent a long afternoon with Damien and me, and then met with Damien and Ryan. We wish she'd been able to see the movie and hope we've done her justice.

As we moved on in the process, Al Worden came on board to help out—his good humor and incredible spirit kept smiles on our faces whenever he was around. Mike Collins was kind enough to take my phone calls; it's been a true privilege to get to know him a bit over the last year or so. Ditto Buzz Aldrin. Gerry Griffin spent hours on the phone with me explaining how Gemini Mission Control worked and helping me sketch out the language—those scenes could not have been written without him. Rick Houston also helped out on the day in Mission Control. And Robert Pearlman read a number of drafts—his encyclopedic knowledge of the program was useful down to the wire.

Of course, I'd be remiss if I didn't again call out Damien, who remains a true inspiration. And Jim, without whom there would be no book. Ryan, Claire, and all the actors contributed to the script, pushing the words forward in ways I couldn't have alone. For that, I am truly grateful.

Isaac Klausner, Wyck Godfrey, and Marty Bowen have been with me for every draft at every step along the way. Isaac knows all of this as well if not better than I do, if there are things missed here then you should blame it on him! Of course, it's due to the hard work of our incredible team that we've managed to recreate these events. Linus Sandgren, Mary Zophres, and Nathan Crowley all did incredible work. Not to mention Paul Lambert, Kevin Elam, J.D. Schwalm, Ian Hunter, Cathy Liu, Erik Osusky, Rory Bruen, Jenne Lee, Scott Robertson, and Adam Merims. And apologies to Tom Cross, Harry Yoon, John To, Derek Drouin, Jennifer Stellema, Jeff Harlacker, and Jason Miller for driving them nuts in post as we tried to get this right.

Capricorn 4 wouldn't be the same without Alissa Goldberg and Erica Tavera who both put in a ton of research. Bette Einbinder and Megan Startz at Universal made this book happen, not to mention Sara Scott, Peter Kramer, and Donna Langley who deftly helped shape the work. Bebe Lerner Baron, Lisa Taback, Katherine Rowe, Paige Phelan, Dustin Thomason, Heather Thomason, and Mike Fisher gave valuable feedback. Ari Greenburg, Jamie Feldman, Jeff Gorin, and Michael Sugar gave wise counsel. And to the extent any of this thinking is good thinking, it's because of everything I've learned from Tommy McCarthy and the *Spotlight* team (ours and the real one); Kristie Macosko Krieger and *The Post* gang; and Steven Spielberg, who's taught me so much, not the least of which is that there is never a bad time for a good cigar.

Finally, in a book about families, I have to thank my own. Grandma Becca, (Grand)Daddio, Aunt Kate, Cousin Bue, Mimi Z, and Pop Pop Andy. And most especially, Laura and Jacob—it's all just empty space and blank pages without you. I love you, as they say, to the Moon and back.

ACKNOWLEDGEMENTS FROM TITAN BOOKS

Titan Books would like to thank everyone who helped with this project: book designer Natalie Clay; the entire cast and crew of *First Man*; Josh Singer for his passion and dedication to this project; and the entire team at Universal Studios, with special thanks to Megan Startz, Bette Einbinder, Dan Abend, and Elizabeth Latham.

ABOVE: Josh Singer, Mark Armstrong, and Jim Hansen.